The Benchmark Lie

How AI-Powered Marketing Makes Experience More Valuable Than Ever

Jeffrey Porzio

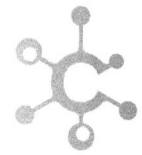

Cadence B2B LLC

Copyright © 2026 by Jeffrey Porzio

All rights reserved.

No portion of this book may be reproduced in any form without written permission from the publisher or author, except as permitted by U.S. copyright law.

This publication is designed to provide accurate and authoritative information in regard to the subject matter covered. It is sold with the understanding that neither the author nor the publisher is engaged in rendering legal, investment, accounting or other professional services. While the publisher and author have used their best efforts in preparing this book, they make no representations or warranties with respect to the accuracy or completeness of the contents of this book and specifically disclaim any implied warranties of merchantability or fitness for a particular purpose. No warranty may be created or extended by sales representatives or written sales materials. The advice and strategies contained herein may not be suitable for your situation. You should consult with a professional when appropriate. Neither the publisher nor the author shall be liable for any loss of profit or any other commercial damages, including but not limited to special, incidental, consequential, personal, or other damages.

Edited by Liz Wheeler

Book Cover by Trent+Co

First edition 2026

Contents

Dedication	VII
Author's Note	IX
Introduction	XVII
1. The $50K Mistake	3
2. Why Averages Lie	9
The Aggregation Fallacy	
The Dimensionality Problem	
Simpson's Paradox	
The False Precision Trap	
Sample Composition Bias	
The Internal Benchmark Illusion	
Why More Data Doesn't Fix This	
How We Got Here	
The Temporal Problem	
The Real Question	
3. The False Comfort of Benchmarks	25
The Illusion of Certainty	
The CYA Economy	
The Unknown Ceiling	
The Streetlight Effect	
The Insight Theater	
The Ecosystem That Won't Let Go	
The Courage to Be Wrong Alone	
The Question We've Been Avoiding	
The Agency Advantage (That Nobody's Actually Using)	

    Bridge to the Alternative

4. Multi-Dimensional Pattern Recognition   49
    The Fundamental Shift
    Why Multi-Dimensional Similarity Works
    The Rhyming Metaphor
    Finding Actionable Insights Benchmarks Can't
    The Five Core Dimensions (Preview)
    How AI Makes This Real
    The Correlation Trap: Why AI Needs Human Judgment
    A Brief Technical Preview
    Why 'Always-On' Intelligence Beats Static Benchmarks

5. How Campaigns Rhyme   71
    Scenario 1: Innovation Seekers in Emerging Categories
    Scenario 2: Challenger Positioning Across Contexts
    Scenario 3: The Hidden Rhyme of Buyer Sophistication
    Scenario 4: When Dimensions Contradict Categories
    What You're Learning to See
    The Uncomfortable Question

6. The Five Dimensions   87
    Dimension 1: Audience Behavior
    Dimension 2: Competitive Environment
    Dimension 3: Message Strategy
    Dimension 4: Execution Approach
    Dimension 5: Business Goals
    Seeing the Five Dimensions in Practice
    How Dimensions Interact: The Reality of Pattern Recognition
    What This Means for You
    The Bridge to What's Next

7. Finding Rhymes—The Human-AI Moat   125
    The Three Phases of AI Value
    Why Off-the-Shelf AI Tools Hit a Ceiling
    The Human-Only Approach
    The Multi-LLM Orchestration Advantage

      Why Institutional Memory Compounds While AI Resets

      Why Quality Control Becomes Harder as AI Gets Better

      The Sustainable Moat

      What This Means for Pattern Recognition

8. How AI Enables Pattern Recognition at Scale   149

      From Strategy to Vectors: The Embedding Breakthrough

      Similarity at Scale: Beyond Simple Matching

      Orchestrating Multiple AI Capabilities

      Persistent Intelligence: How Pattern Knowledge Compounds

      Transparency and Auditability: Not a Black Box

      The Orchestra Conductor Metaphor

9. Privacy-First Intelligence   159

      What Pattern Recognition Actually Requires

      Why Campaign Specifics Are Actually Noise

      Architecture That Can't Compromise What It Doesn't Capture

      The Two-Layer Privacy Framework

      The Ethical Framework

10. Evaluating Pattern Recognition Approaches   173

      The AI Washing Problem

      Red Flags—What to Avoid

      Green Flags—What Signals Real Capability

      Questions That Separate Signal from Noise

      Why Most Vendors Will Fail These Tests

      The Bottom Line

11. The Choice Ahead   199

      Path 1: Building Internal Capability

      Path 2: Partnering with Specialists

      Path 3: Starting with Dimensional Thinking

      The Decision Framework

      The First-Mover Opportunity

      The Scarcity Reality

      The Platform AI Trap

      A Note on Evaluation

The Choice Is Yours

A Final Word 211

For my sons,
who already know the best answers
come from asking better questions.

Author's Note
Why I Wrote This Book

THE MOMENT I COULDN'T unsee it involved two campaigns that had nothing to do with each other.

One targeted chief compliance officers at mid-market insurance companies. The other targeted IT decision-makers at large retailers. Different industries. Different titles. Different products. Different team members working on them. By every conventional measure, these campaigns shouldn't have been comparable.

Yet they performed identically well, following the same engagement patterns, the same conversion curves, and the same timing of decision triggers. When I dug deeper, the reason became clear: their *dimensions* rhymed. Both targeted risk-averse buyers who needed to build internal consensus before making purchasing decisions. Both faced intense competitive pressure from entrenched incumbents with deeper pockets. Both were selling complex solutions requiring behavioral change, not just technology adoption. Both audiences were sophisticated enough to smell generic marketing from a mile away.

No industry benchmark would have connected those dots. "Insurance" and "retail" live in separate categories. The campaigns would never appear in the same report, never get averaged into the same metric. But they were teaching the same lesson.

And I'd been missing lessons like this for 25 years.

I've led digital marketing teams for some of the largest brands in the world. I've worked across thousands of campaigns in partner ecosystems, enterprise software, cloud platforms, consumer packaged goods, financial services, and business-to-business (B2B) services. I watched search engine optimization (SEO) emerge from nothing, social media transform from novelty to necessity, and Flash disappear into irrelevance. I've seen paradigm shifts come and go.

But one thing remained constant: the gnawing sense that something fundamental was missing from how we measured success.

I've been in the room when agency teams scramble after learning that a campaign we celebrated as "significantly outperforming benchmarks" had zero impact on the client's business. The panic is real. The blame loops are exhausting. Marketing points to engagement metrics. Sales points to lead quality. The agency points to industry standards. Everyone has data proving they did their job correctly.

And nobody asks the question that actually matters: *What worked to drive business outcomes, and why?*

That's not quite true. The question was being asked constantly. But it was always being answered in silos. "What's working to drive clicks on our LinkedIn ads?" "Which email subject lines get higher opens?" "What display creative generates the most engagement?"

Nobody was asking how LinkedIn impacted integrated campaigns designed to drive business revenue. Nobody was connecting the dimensions that actually predicted whether prospects would convert six months later. We were optimizing individual channels against their own benchmarks while the business outcomes remained disconnected from our measurement systems.

Here's a truth agencies don't say out loud: benchmarks are incredibly easy to beat.

I can count on one hand the number of campaigns I've touched in the past five years that performed below benchmark. Maybe five out of hundreds. Everything else "outperformed industry standards." Which should have been a red flag that the standards themselves were meaningless.

But beating benchmarks became the goal because it was achievable, measurable, and politically safe. You could walk into a client meeting with green dashboards and confident narratives about success, even when the sales team was quietly struggling with leads that never converted.

AUTHOR'S NOTE

The measurement system rewarded the wrong outcomes. And we all participated in it because there wasn't an alternative framework.

Testing presented another set of impossible constraints.

Agencies operate primarily on time-and-materials models. If I wanted to test a new message hypothesis, I needed either a client willing to spend incremental funds or my agency to be willing to engage copywriters and designers at our own cost. Neither option was particularly welcome.

Clients would say things like, "Well, you're the experts. You do this all the time. You should know what the right messages are for a campaign like this." They often failed to recognize that our target audience wasn't "one person," despite how marketing personas made them seem.

So agencies adapted by creating all messaging vehicles upfront. Everything got neatly packaged for client review and approval, then went to market. When optimizations happened, they focused on media buying and targeting because re-engaging creative teams cost money. The messaging itself remained static, approved and frozen, regardless of what performance data revealed about what actually resonated.

We tested media placements endlessly. We rarely tested the messages themselves. The incentive structure made genuine experimentation financially painful.

For years, I knew there was *something* missing. I could sense patterns that industry categories couldn't capture. I'd see campaigns succeed or fail for reasons that had nothing to do with their benchmark comparisons. But I lacked the framework to articulate it, and more critically, I lacked the analytical capacity to test what was possible.

There simply weren't enough people or spreadsheets to analyze campaigns across dozens of dimensions simultaneously. The human brain can track maybe three to five variables before analysis paralysis sets in. Manual pattern recognition at that scale was functionally impossible.

Until AI became commercially viable.

When large language models (LLMs) and multi-dimensional analysis became accessible, everything clicked. Suddenly, the pattern recognition I'd been doing intuitively for decades could be systematized, scaled, and validated. I could finally test the hypothesis I'd been carrying: that campaigns rhyme across dimensions that matter, not within categories that don't.

But here's what I've also witnessed: agencies and brands struggling to implement AI into their processes and their people. I've seen the same mistakes repeated, treating AI as a replacement for human judgment rather than an amplification of it. I've watched teams feed performance data into AI tools and get back sophisticated-sounding nonsense because nobody with deep expertise was guiding what to look for.

AI doesn't inherently know that message sophistication matters more than industry category. It doesn't know which dimensional combinations predict success versus which are statistical noise. It doesn't understand buyer psychology or competitive dynamics or organizational decision-making in ways that decades of experience teach you.

That knowledge comes from humans. From 25 years of watching campaigns succeed and fail. From recognizing patterns that aren't obvious in the data but become clear when you've lived through thousands of similar situations.

This is the pattern I kept encountering: AI makes sophisticated pattern recognition possible at scale, but only when directed by experienced human judgment. Neither works without the other.

Which brings me to how this book was written.

I wrote *The Benchmark Lie* in partnership with Claude, an AI assistant from Anthropic.

What I brought to this partnership: the complete intellectual framework, the theories, the 25 years of accumulated pattern recognition across thousands of real campaigns, the stories drawn from direct experience, and the strategic insights that can only come from living through agency panics and client struggles and competitive battles. Every idea in this book originated from my expertise. Every argument reflects patterns I recognized long before AI existed. Every framework represents my theory about how marketing intelligence should actually work.

What do I mean by "rhyming"? I chose that metaphor deliberately. "Related campaigns" implied rigid connections, solid lines between similar things. That's how everyone was already analyzing campaigns, drawing straight lines between matching categories. But what I was seeing wasn't relatedness. It was dimensional resonance. As Eminem demonstrated in that famous *60 Minutes* interview, rhyming isn't about identical words; it's about creative pattern recognition across phonetic dimensions. "Orange" rhymes with "door hinge" if you're willing to think dimensionally rather than categorically. Campaigns work the same way. Insurance compliance and retail IT monitoring don't "relate," but their dimensions rhyme. You'll see this concept explored in depth in Chapter 4.

Here's what Claude provided: the ability to articulate those ideas with clarity and structure at speed. Partnership in refining arguments. Help organizing complex thinking into a coherent narrative. The capacity to draft and iterate rapidly, turning decades of accumulated insight into systematic explanation faster than would have been possible working alone.

But here's what matters most about how I used this tool.

There's a misconception that AI assistance means passive order-taking. You might think that I gave general instructions, and Claude did most of the work. That fundamentally misunderstands how I chose to use this technology.

I specifically trained Claude to function as a collaborative partner, not a compliant assistant. I directed it to challenge my assumptions, question my framing, and push back on ideas that didn't hold up to scrutiny. Throughout this process, sometimes I convinced Claude my approach was right. Sometimes Claude convinced me there was a better way to structure an argument. Sometimes we discarded ideas entirely because neither of us could make them work.

That's intentional collaboration. I've always thrived in environments where ideas get stress-tested, where disagreement strengthens thinking, where the best solution emerges from rigorous back-and-forth rather than unquestioned acceptance. I engineered that same dynamic with AI.

The difference with AI collaboration isn't that the AI tools do all the work. It's the speed at which you can iterate through multiple perspectives and land on the strongest version of an idea. You can be generative, critical, and synthetic in rapid cycles that would take weeks with traditional human collaboration or solo work.

Speed matters. Not just for efficiency, but for maintaining momentum while ideas are fresh, for testing multiple approaches before settling on the best one, and for getting sophisticated thinking to market while it's still relevant.

I'm transparent about this AI partnership for two reasons.

First, because the entire thesis of this book is that human expertise combined with AI capabilities creates a compounding competitive advantage. It would be intellectually dishonest to advocate for that partnership while hiding that I used exactly that approach to write it.

The "meta" nature of this wasn't lost on me: I'm arguing that dimensional pattern recognition requires humans and AI working together, and I'm demonstrating that partnership in real time by writing about it.

Second, because I believe transparency matters. Some people will dismiss this book because AI was involved in its creation. That's fine. They're self-selecting out of the conversation about how knowledge work actually evolves. Others will understand that AI is a tool, a remarkably sophisticated one, that amplifies human capability rather than replacing it.

Every framework in this book is mine. Every argument draws from real experience. Every insight reflects patterns I've recognized. Claude didn't invent dimensional pattern recognition or discover that benchmarks fail mathematically or develop the five-dimension framework. I created all of that through decades of observation, experimentation, and hard-won understanding.

What Claude provided was partnership in articulation and speed to market. The ability to structure complex ideas clearly. The capacity to iterate rapidly. Exactly the kind of human-AI collaboration I'm advocating for throughout these pages.

If that bothers you, this probably isn't your book.

AUTHOR'S NOTE

If it intrigues you, if you're curious about what becomes possible when experienced judgment directs sophisticated AI capabilities, then keep reading.

This book challenges fundamental assumptions about how digital marketing is measured and optimized. It argues that industry benchmarks aren't just imperfect, they're actively harmful, creating expensive illusions of success while hiding genuine opportunities.

The alternative I'm proposing, multi-dimensional pattern recognition, requires both technological capability and human expertise. It requires understanding that campaigns don't repeat, but their dimensions rhyme. It requires recognizing that the insights hiding in your data aren't visible through traditional category-based analysis.

And it requires accepting that the future of marketing intelligence isn't humans *or* AI. It's humans *and* AI, working together in ways that amplify what each does best.

I wrote this book for agency principals who sense their cross-client experience should provide more value than it currently does. For marketing leaders in B2B and beyond who've suspected that "beating the benchmark" is a hollow victory. For strategists and consultants who've questioned conventional wisdom but lacked the framework to articulate what's actually broken.

For anyone who's been in that conference room when the metrics say success but the business says failure, and wondered: *What are we missing?*

You're not missing better benchmarks. You're missing a completely different way of thinking about campaign intelligence.

That's what this book provides.

Let's begin.

Jeffrey Porzio
Boston, Massachusetts
November 2025

Introduction

THIS BOOK IS WRITTEN for experienced marketers who already sense something's wrong with how our industry measures success. If you've ever stared at a benchmark report knowing it was meaningless but presented it anyway because that's what boards expect, this book is for you. Here's a guide describing what the book is and isn't, who it's intended for, and how to read it:

What this book is

An intellectual argument for why industry benchmarks fail mathematically and what should replace them. It's strategic thinking about campaign intelligence, not tactical instruction about implementation.

What this book isn't

A how-to guide with worksheets and templates. You won't find step-by-step instructions for building pattern recognition systems or DIY frameworks for dimensional analysis. That's deliberate.

Who should read this book:

- Chief marketing officers (CMOs) and marketing leaders frustrated with benchmark theater
- Agency principals trying to differentiate in an AI-saturated market
- Marketing strategists who've always known benchmarks were bullshit
- Consultants seeking frameworks for more sophisticated campaign analysis

- Anyone evaluating AI-powered marketing intelligence claims

How to read the book

The book builds progressively. Part 1 establishes why benchmarks fail—emotionally, mathematically, and systemically. Part 2 introduces dimensional pattern recognition as an alternative. Part 3 explores implementation realities. Each section assumes you've absorbed what came before.

You could skip directly to Part 2 if you already know benchmarks are broken. But Part 1 provides the intellectual foundation that makes the alternative compelling rather than just interesting.

A note on examples

Throughout this book, I use realistic campaign scenarios to illustrate dimensional patterns. These aren't case studies with specific client names and performance metrics—those details would violate the privacy principles this approach is built on. They're composite examples drawn from 25+ years of experience, designed to show how dimensional thinking works without revealing proprietary information.

A note on technology

This book discusses AI extensively, but it isn't only about AI. It's about what becomes possible when experienced marketers orchestrate AI capabilities for sophisticated pattern recognition. The technology enables the approach but doesn't define it.

Ready? Let's begin with a story about a $50K mistake that looked like success.

PART 1
THE PROBLEM

Chapter One

The $50K Mistake

Somewhere right now, a marketing team is celebrating metrics that are quietly destroying their business.

They're hitting benchmarks. They're beating industry averages. Their dashboards glow green with quarter-over-quarter improvements in opens, clicks, and engagement. Executive leadership sees the numbers and approves increased budgets. Agencies point to the same data and renew contracts. Everyone agrees: the campaigns are working.

Except they're not. Sales teams watch conversion rates collapse. Revenue targets slip. Competitive deals are lost to companies running similar campaigns to similar audiences—but getting radically different results. And the entire organization burns weeks in circular blame loops, each stakeholder armed with their own dashboard proving they did their job correctly.

The culprit isn't poor execution or a bad creative team or lazy salespeople. It's the measurement system itself. When you optimize campaigns against generic industry benchmarks—statistical averages that blend together thousands of incomparable efforts—you create a dangerous illusion of success. You're measuring yourself against everyone, which means you're measuring yourself against no one.

Here's how it typically unfolds.

The dashboard was green.

Sarah Chen, CMO of a fast-growing B2B software-as-a-service (SaaS) company, advanced to the next slide in her quarterly board presentation. Email performance metrics filled the screen—a 26% open rate and a 3.2% click-through rate, both tracked in tidy upward arrows from the previous quarter.

"Our financial services lead generation campaign is performing exceptionally well," she said, voice steady with confidence. "We're not just hitting our targets—we're outperforming B2B industry benchmarks by double digits. Open rates are up 18% quarter-over-quarter, and click rates are up 22%."

The board nodded. One director leaned forward: "And the benchmark comparison?"

Sarah clicked to the next slide. A clean chart showed their performance hovering comfortably above a horizontal line labeled "B2B Industry Average." Green everywhere.

"We're consistently beating industry standards. The campaign generated 847 new leads from financial services executives, and our nurture sequences are engaging them at rates well above what we'd typically expect."

Budget discussions moved forward. The $50K campaign had delivered exactly what it promised: measurable success against external benchmarks, quarter-over-quarter growth, and a healthy pipeline of leads.

Sarah closed her laptop, feeling vindicated. Marketing was doing its job.

Three weeks later, the cracks started showing.

Dante Russo, VP of business development, pulled up his dashboard on a call with Sarah. Unlike marketing's green indicators, his showed red.

"Financial services' lead-to-opportunity conversion is down 34% year-over-year," he said, showing a trend line in consistent decline. "Quarter-over-quarter, we're down another 18%. We're sounding the alarm because this isn't stabilizing—it's accelerating."

Sarah's stomach tightened. "But engagement is up. Opens, clicks—"

"My reps are nurturing these leads for weeks and getting nowhere," Dante interrupted. "Out of 847 leads, we've converted eleven opportunities. Eleven. Last year, same campaign, same vertical? We were at 4.7% conversion. We're down two-thirds."

The agency meeting a few days later started defensively.

"Your email performance is exceptional," said Roman, the agency's director of digital strategy. "You're crushing B2B industry benchmarks. Everything is performing above target. In fact, our subject line testing is one of the reasons you're seeing that steady increase in open rates quarter-over-quarter."

"Sales is saying these leads are trash," Sarah countered.

The word hung in the air.

Roman's jaw tightened. "We delivered 847 qualified leads from financial services executives—verified titles, confirmed intent signals. We've nurtured them with sequences significantly outperforming industry standards."

Amanda, Sarah's brand marketing director, jumped in: "Let's talk about these benchmarks. Where exactly do they come from? Because we've been beating them for three quarters straight, and somehow revenue is declining."

"They're industry-standard metrics," Roman said. "B2B averages aggregated across—"

"Is B2B even an industry?" Amanda interrupted. "What are we doing to personalize these emails?"

"We're personalizing. First name, company name, industry-specific messaging—"

Amanda pulled up one of the nurture emails. "'Hi [First Name], As a leader at [Company Name] in the Financial Services industry...' That's not personalization, Roman. That's mail merge."

Roman's tone shifted, careful now. "Well, as a reminder, we presented all creative back on September 29th, and that reflected the personalization strategy we discussed. We received email approval for all creative on October 1st."

The Zoom call devolved into a circular blame loop.

Behind the scenes, the agency panicked.

Roman summoned his team: senior account leads, digital execution, and the creative director. Everyone crammed into a conference room, with stress levels climbing.

"Our targeting was solid," the digital lead said. "Intent signals, verified titles, firmographic match."

"Maybe they're just taking longer to convert?" someone suggested. "Financial services has longer sales cycles."

"Or maybe it's their nurture," the creative director offered. "Once leads leave our sequences and sales takes over, who knows what messages they're sending?"

"Here's what we do," Roman said, landing on a plan. "Next campaign, we segment by sub-industry. Retail banking gets different messaging than wealth management, and insurance gets its own track. That's more sophisticated targeting."

Heads nodded around the table. It seemed logical. More segmentation, more precision.

Nobody noticed the flaw: they were just creating more groups to treat as groups. More buckets, same problem. And nobody considered that an insight from how insurance executives responded might actually help lift performance with consumer banking executives—that patterns might rhyme across sub-industries in ways their categories would never reveal.

They spent the next hour crafting a response that meticulously documented their success metrics while gently suggesting the issue lay downstream in the client's sales process.

Nobody asked: What made the eleven conversions work?

The pressure intensified.

Amanda started quietly reaching out to other agencies. Dante's business development representatives (BDRs) kept burning cycles on leads that went nowhere. Competitors closed deals with the same executives who'd engaged with Sarah's emails but never converted.

And everyone—Sarah, Amanda, Roman, Dante—treated the 847 leads as a single mass: "the financial services leads." And they saw one verdict: *not converting*.

Nobody noticed that the eleven successful conversions shared something in common. Nobody asked whether retail banking executives behaved differently from wealth management chief financial officers, or whether financial technology (fintech) founders responded to different messages than insurance executives. Nobody considered that "financial services" wasn't an industry—it was a lazy label slapped onto wildly different companies with completely different needs.

The agency's solution—segment by sub-industry—would just recreate the same problem at a smaller scale. They'd optimize retail banking emails against retail banking benchmarks, wealth management against wealth management. More categories. Same statistical fiction.

The human brain can only track so many dimensions before analysis paralysis sets in. Three segments? Manageable. Five? Stretching. Ten sub-industries across four title levels across three intent topics? Impossible. So they simplified. They bucketed. They averaged.

And they missed the patterns hiding in plain sight.

The $50K mistake wasn't the media spend, or the execution, or the creative work.

It was the measurement.

They'd optimized against "B2B industry benchmarks"—a statistical fiction averaging email performance across thousands of incomparable campaigns. They celebrated beating a number with no connection to their actual business goals. They compared themselves to everyone and therefore to no one.

While they argued about responsibility, the real question sat unasked:

What campaigns actually rhyme with ours? Not "B2B" campaigns—but campaigns targeting similar buyer behaviors, similar competitive contexts, similar decision-making processes in similar organizational structures.

They didn't have that framework. They didn't have that language. They didn't even know it was possible.

So they did what most marketing teams do: they fired the agency, hired a new one, and prepared to make the same mistake again.

Nobody was asking why some leads converted while others didn't.

And that's where the real opportunity was hiding.

Chapter Two

Why Averages Lie

WHICH U.S. CITY HAS the best restaurants?

Let's say some survey tells you the average restaurant rating in New York is 3.8 stars, while Chicago averages 3.6 stars. New York wins, right?

Except that 3.8 blends together the Michelin three-star tasting menu in Midtown, the airport Sbarro in LaGuardia, the corner bodega selling day-old sandwiches, and the hotel breakfast buffet in Times Square. They're all "restaurants in New York." None are comparable. And that 0.2-star difference? It might just mean New York has more hotel buffets dragging down the average, or Chicago has stricter rating criteria, or tourists in New York are more generous with stars.

If you're planning where to travel for an anniversary dinner, that average tells you nothing. Worse—it creates the illusion you're making an informed comparison, when you're just measuring noise.

The only restaurants that 3.8 average describes are the ones that don't exist: the perfectly average restaurant serving the perfectly average meal to the perfectly average diner.

You'd never plan a trip based on "average city restaurant rating." You'd ask: *Which city has the kind of restaurants I'm looking for?*

Yet marketers routinely evaluate campaigns by asking "What's the average open rate for B2B SaaS?" instead of "What campaigns are actually comparable to mine?"

The math isn't complicated. The lie is.

The Aggregation Fallacy

Every industry benchmark suffers from the same fundamental problem: it averages things that shouldn't be averaged.

Take that "22% average open rate for B2B SaaS" that Sarah's team celebrated in Chapter 1. What actually goes into that number?

A seed-stage startup announcing their product launch to 500 warm leads who signed up last week. A mid-market vendor sending quarterly newsletters to 50,000 contacts acquired over five years. An enterprise platform distributing technical documentation to existing customers. A struggling competitor blasting promotional offers to a purchased list. A category leader sending investor updates to their board and shareholders.

Oh, and that average also blends together tools that cost $49/month with platforms requiring $200,000 annual commitments. Because apparently a productivity app targeting solopreneurs and an enterprise data infrastructure platform are comparable as long as they're both "SaaS."

All B2B. All SaaS. All email. None comparable.

This is the *aggregation fallacy*: combining measurements across incomparable situations and pretending the result means something. Statisticians call it "comparing apples and oranges." But that's too generous. We're comparing apples and hub caps and the concept of Thursday.

Here's what makes it particularly insidious: the math is correct. If you collect 10,000 B2B SaaS email open rates and calculate the mean, you'll get a number. That number is mathematically accurate. You can verify the calculation. You can put it in a slide deck. You can compare it quarter over quarter and watch it trend up or down.

But accuracy and usefulness are not the same thing.

The average temperature across all U.S. cities today is 61°F. Accurate? Yes. Useful for deciding what to wear in Minneapolis in January? No. Because Minneapolis isn't experiencing "average America"—it's experiencing Minneapolis.

Your campaign isn't experiencing "average B2B SaaS." It's experiencing your specific audience, your specific message, your specific competitive context, your specific company stage, your specific deliverability reputation, your specific subject line approach, and dozens of other variables that the benchmark erased to create that clean 22% number.

The benchmark describes a campaign that doesn't exist, run by a company that doesn't exist, targeting an audience that doesn't exist.

And you're optimizing against it.

The Dimensionality Problem

Let's say you accept that not all B2B SaaS campaigns are comparable. Fine—you'll get more specific. You'll segment by company size. Or industry vertical. Or job title. Or region.

This is what Sarah's agency tried. Remember their solution? Segment financial services by sub-industry. Retail banking gets its own benchmark. Wealth management gets another. Problem solved.

Except it's not.

Because campaign performance doesn't vary across one dimension, or five, or even ten. It varies across dozens of dimensions simultaneously—and most of them aren't captured in any benchmark database.

Here are just some of the variables that meaningfully impact email performance:

Audience dimensions

Company size, industry, job title, seniority, department, geography, buying stage, previous engagement history, competitive product usage, organizational structure, decision-making authority, budget ownership, technical sophistication, risk tolerance

Message dimensions

Value proposition complexity, competitive positioning, call-to-action specificity, subject line style, personalization depth, content format, reading level, urgency framing, social proof inclusion, visual design approach

Context dimensions

Market maturity, competitive intensity, regulatory environment, economic conditions, seasonal timing, news cycle relevance, industry trends, technology adoption curves, hiring patterns, mergers and acquisitions (M&A) activity

Sender dimensions

Brand recognition, domain reputation, sending consistency, list hygiene, authentication protocols, previous campaign performance, unsubscribe rates, complaint rates, engagement patterns, sender personality

That's 60+ variables. And I'm simplifying.

And here's something most marketers never consider: even metrics that seem straightforward are measuring different things across different campaigns. Your "unsubscribe rate" might look great partly because emails flagged by corporate firewalls never reach an inbox at all—the recipient never saw your message, never had the chance to unsubscribe, so your rate stays artificially low. Meanwhile, those "delivered" emails aren't contributing to opens, clicks, or conversions. You're comparing your measurement of one thing against someone else's measurement of something slightly different, calling them the same metric, and averaging them together.

Now here's the problem: when you create a benchmark, you're collapsing all of these dimensions down to a handful of categories. "B2B SaaS" captures maybe three of those 60+ variables. "B2B SaaS to mid-market CFOs" might capture eight. "B2B SaaS to mid-market CFOs in financial services" might get you to twelve.

You're still ignoring 48+ dimensions that actually matter.

It's like rating restaurants based only on "type of cuisine"—ignoring price point, location, ambiance, chef experience, ingredient sourcing, service quality, noise level, parking availability, reservation difficulty, and everything else that determines whether you'll enjoy your meal.

Categories are real. They're just not sufficient.

And here's what makes this particularly frustrating: the more dimensions you add to your segmentation, the smaller your sample size becomes. Eventually, you're comparing yourself to three other campaigns, and you've lost statistical significance entirely. You can't win. Either your benchmark is too broad to be useful, or too narrow to be reliable.

The fundamental approach—categorize, then average—is mathematically doomed.

Simpson's Paradox

But surely if the overall number is good, that's still meaningful? Even if the benchmark is imperfect, beating it consistently must indicate something positive?

Not necessarily. Welcome to Simpson's Paradox.

Simpson's Paradox occurs when a trend appears in aggregated data but reverses when you examine the underlying groups. The phenomenon is named after Edward H. Simpson, who described it in a technical paper in 1951. It's one of the most counterintuitive statistical phenomena, and it happens constantly in marketing benchmarks.

Here's how it plays out in email:

Your overall email open rate: 24%. Industry benchmark: 22%. You're winning!

But when you segment by actual audience behavior, you'll see this:

High-intent prospects (actively evaluating solutions): Your performance is 31%; comparable campaigns 42%

Medium-intent contacts (generally aware of problem): Your performance is 18%; comparable campaigns 28%

Low-intent subscribers (opted in but not currently buying): Your performance is 8%; comparable campaigns 11%

You're losing in every meaningful segment. But because you have more low-intent subscribers than your competitors (maybe you're better at top-of-funnel lead capture), your overall average looks great. The aggregate number is hiding systematic underperformance.

Or take display advertising. Your banner campaign achieves a 0.18% click-through rate. The industry benchmark is 0.12%. Victory!

Until you realize you're running primarily on low-quality programmatic inventory where accidental clicks are common, while competitors are running on premium publisher sites where every click represents genuine interest. Your higher click-through rate (CTR) is actually a warning sign—you're paying for worthless clicks—but the benchmark makes it look like success.

This isn't a theoretical edge case. This is Tuesday for most marketing teams.

Sarah's "successful" campaign in Chapter 1? They were probably experiencing exactly this. The overall numbers looked great. But if you segment by the actual buying stage, actual competitive context, and actual message relevance, the story might have been very different.

Simpson's Paradox reveals why "beating the benchmark" can actually correlate with losing deals. You're not measuring the right thing. You're measuring an aggregate that masks what's actually happening in the segments that matter.

The benchmark doesn't lie about the math. It lies about what the math means.

The False Precision Trap

Let's talk about those decimal points.

"Industry open rates increased from 22.1% to 22.3% quarter over quarter, indicating improving engagement."

No. It doesn't indicate that. It indicates that someone calculated two numbers and subtracted them.

Or how about this: "Average video completion rate for B2B content is 47.8%, up from 46.9% last quarter."

Really? You measured video viewing behavior across thousands of campaigns—some auto-playing on mute, some requiring clicks to start, some on platforms that count "completion" at 50% viewed, others at 95%—and you're confident that a 0.9% difference is meaningful?

When you're averaging together thousands of incomparable campaigns, those decimal points aren't precision—they're noise dressed up as signal.

Imagine trying to measure the average depth of the ocean by having a thousand people dip rulers into random locations and report back their findings. Some are measuring tide pools. Some are measuring the Mariana Trench. Most are measuring something in between. You average all the numbers together and report: "Average ocean depth: 3,688.2 meters."

That .2 at the end—does it mean anything? Does it represent real measurement precision? Or is it just a mathematical artifact from averaging measurements that shouldn't have been averaged?

In campaign benchmarks, it's the latter.

The 0.2% difference between 22.1% and 22.3% is almost certainly smaller than the measurement error in the underlying data. Different email clients report opens differently. Different tracking pixels fire at different times. Different audiences have different privacy settings. Different senders have different list hygiene practices. Different segments have different engagement patterns.

All of that variance—all that real, meaningful, systematic difference in what's actually being measured—is far larger than 0.2%.

But that decimal point creates an illusion of scientific rigor. It suggests someone measured carefully, controlled for variables, and arrived at a precise conclusion. In reality, someone aggregated messy data from incomparable sources and calculated a mean. The precision is false. The confidence is misplaced.

For example, if a benchmark report tells you the industry average is 22.3%, ask yourself: "What would need to be true for that .3 to be meaningful?"

Answer: Every campaign in the average would need to be measuring the same thing, in the same way, with the same level of accuracy, across the same time period, to the same type of audience, with comparable definitions of success.

That has never happened. That will never happen.

The decimal points are theater.

Sample Composition Bias

But at least when you have enough data, the errors should average out, right? The more campaigns you include, the more reliable the benchmark becomes?

Only if the sample is representative. And it never is.

Benchmark data comes from somewhere. Usually it's coming from one of three sources:

Vendor customers

If you're looking at HubSpot's benchmark report, you're seeing data from companies that chose HubSpot. That's already a non-random sample—they're companies with certain budgets, certain technical sophistication, certain marketing maturity, certain industry focus. Their performance doesn't represent "average marketing." It represents "marketing done by companies that chose HubSpot."

Survey respondents

If it's based on a survey, you're seeing data from people who chose to respond. Who responds to marketing performance surveys? Usually people with something to prove. High performers who want to share their success. Or struggling performers looking for solutions. Average performers are busy running campaigns—they're not filling out surveys.

Aggregator platforms

If it's from a data aggregator or analytics platform, you're seeing whoever integrates with that platform and agrees to share data. That's a self-selected sample with unknown biases.

None of these samples is representative of "the industry." They're representative of "companies that report data to this particular source."

And here's where it gets worse: the sample composition changes over time in non-random ways. During economic booms, more companies run campaigns—the sample gets diluted with inexperienced marketers. During recessions, struggling companies cut marketing budgets—the sample skews toward better-funded, higher-performing teams. The benchmark moves, but not because marketing performance changed. It moved because the sample changed.

You have no way to know what you're actually being compared against.

It's like trying to determine if you're a good driver by comparing your accident rate to "other drivers"—but the comparison group is limited to people who voluntarily reported their driving record to a particular insurance website. That's not "other drivers." That's "other drivers who use this website and chose to share their data." Completely different population.

The benchmark isn't measuring the industry. It's measuring whoever happened to show up in the data collection process.

And you're optimizing against it.

The Internal Benchmark Illusion

Some marketers recognize that industry benchmarks are too broad, so they create their own. They average their company's historical campaign performance—all their email CTRs, all their display click-throughs, all their video completion rates—and use that as their benchmark.

"We don't compare to industry averages," they say. "We compare to our own performance. Much more reliable."

Except it's not. It's the same math problem with a smaller dataset.

Your brand's "average email open rate" still blends together all of the following:

- Product launch announcements to hot prospects

- Weekly newsletters to cold subscribers

- Event invitations to regional segments

- Customer onboarding sequences

- Sales follow-ups to demo attendees

- Quarterly updates to dormant contacts

All your campaigns. Still incomparable.

You've just moved from comparing yourself to everyone's incomparable campaigns to comparing yourself to your own incomparable campaigns. The aggregation fallacy doesn't care about sample size or data ownership—it cares about whether the things being averaged should be averaged.

In some ways, internal benchmarks are more dangerous because they *feel* scientific. You controlled the data collection. You know the source. You trust the measurement. But you're still averaging apples and hub caps and the concept of Thursday—they just all happen to be *your* apples, hub caps, and Thursdays.

The illusion of control masks the fundamental problem: you're still asking the wrong question.

Even if the average were calculated correctly from a representative sample, it still wouldn't tell you what you need to know. Because averages hide distribution shape.

"Average email CTR is 3.2%" could mean either of the following two things:

Scenario A: Everyone gets between 2.8% and 3.6%. Tight clustering around the mean. Consistent performance across the sample.

Scenario B: Half get below 1%. Half get above 6%. Massive variance. The average describes almost no one.

Which scenario are you in? The benchmark doesn't tell you.

This matters enormously for decision-making. If you're getting a 3% CTR in Scenario A, you're average—middle of the pack. If you're getting a 3% CTR in Scenario B, you might be in the bottom quartile, with massive room for improvement.

But all you see is "3.2% is the industry average." You have no idea if you're in a tight race or getting lapped.

Statisticians solve this by reporting not just the mean, but also the standard deviation, the median, the quartiles—the shape of the distribution. Benchmark reports almost never include this. They give you the average and call it a day.

Why? Because showing the distribution would reveal how meaningless the average is. It would show that campaigns are scattered all over the map, with no meaningful center. It would show that "average" describes a campaign that doesn't exist.

So they hide the variance and report the mean. It looks cleaner. It's easier to understand. And it's useless.

You're flying blind, pretending the altimeter is working.

Why More Data Doesn't Fix This

The natural response to all these problems is: "Fine, we need better benchmarks. More data, more segmentation, more sophisticated analysis."

No. More data doesn't fix bad methodology.

If you're averaging incomparable things, averaging 100,000 of them doesn't make the result more useful than averaging 1,000. It just means you're computing a more precise measurement of something meaningless.

This is the "garbage in, garbage out" principle at scale. Adding more garbage doesn't create gold. It creates a larger pile of garbage with more decimal points.

In fact, more data can make the problem worse by creating false confidence. When a benchmark report says "Based on analysis of 250,000 campaigns," that sounds authoritative. Scientific. Reliable. It suggests someone did serious research and arrived at solid conclusions.

But if those 250,000 campaigns are incomparable—if they're mixing together product launches and quarterly newsletters and sales promotions and customer onboarding and abandoned cart reminders—then the sample size is irrelevant. You've just computed the average with more precision. You haven't computed anything meaningful.

This might be counterintuitive, but 10 genuinely comparable campaigns would give you more useful intelligence than 100,000 incomparable ones.

But benchmarks don't do that. They can't do that. Because creating genuinely comparable samples would require evaluating campaigns across dozens of dimensions simultaneously—and at that point, you wouldn't be creating benchmarks anymore. You'd be doing pattern matching.

Which is exactly where we're headed. But not yet.

How We Got Here

Before we move forward, it's worth asking this: if benchmarks are this broken, why do they exist?

The cynical answer: because someone needed something to sell.

Marketing data aggregators and analytics platforms collect campaign performance data as a byproduct of their core business. They're sitting on mountains of email opens, click rates, and conversion metrics. That data has value—but only if they can package it into something marketable.

Enter benchmarks.

Take all that data, average it by industry, and suddenly you have a "report" you can put behind a form. You can generate press releases: "Email Open Rates Declined 3% in Q3." You can create annual studies. You can position yourself as a thought leader with unique data insights.

The fact that the benchmarks are statistically meaningless doesn't matter. They look scientific. They're based on "real data." They give marketers something to put in their presentations to justify their budgets.

Everyone wins. The aggregator gets leads. The marketer gets political cover. The executive gets a number to track quarter over quarter.

The only loser is the campaign performance—but that's harder to measure, so no one notices.

This isn't a conspiracy. It's just economics. Data aggregators aren't evil—they're responding to market demand. Marketers want benchmarks. Executives demand them. So vendors provide them. The fact that they're providing something that actively misleads their customers is... well, let's call it an uncomfortable externality.

There's also a more generous explanation: most people genuinely don't understand how badly aggregation fails at this scale. Statistics courses teach you about means and medians and standard deviations, but they rarely teach you about what happens when you average across dozens of uncontrolled dimensions. It's technically correct math applied to situations where the math doesn't help.

So you get well-meaning analysts creating well-meaning reports that are mathematically accurate and practically useless.

Either way, the result is the same: an industry built on comparing campaigns to statistical fictions.

And the longer it continues, the more expensive the lie becomes.

The Temporal Problem

There's one more issue worth naming before we close this chapter: benchmarks are always backward-looking.

"2024 email performance averaged 22% open rates." Great. What does that tell you about your campaign launching in Q1 2025?

Maybe nothing. Because the market has shifted.

Apple's Mail Privacy Protection changed how opens are tracked. Gmail reorganized its inbox categories. Microsoft changed its spam filtering algorithms. LinkedIn adjusted its algorithm. A major competitor launched an aggressive campaign that saturated your audience. Regulatory changes affected how you can target prospects. Economic conditions shifted buyer priorities.

All of that happened between when the benchmark data was collected and when you're running your campaign. The benchmark describes a past that no longer exists.

Marketing is not a stable system. It's a constantly evolving competitive landscape where yesterday's tactics become today's noise. Buyer expectations shift. Channel effectiveness changes. What worked in aggregate last year might fail this quarter—not because you executed poorly, but because the ground moved.

Benchmarks give you a rearview mirror when you need a forward-looking radar.

And even if the market were stable, your campaign is unique to this moment. Your competitive context today is different from the aggregate of everyone else's competitive contexts last quarter. Your message right now is responding to current events, current buyer mindsets, and current market dynamics.

The benchmark can't see any of that. It can only tell you what happened, on average, to a bunch of incomparable campaigns, in the past.

You're making forward-looking decisions based on backward-looking aggregates of situations that were never comparable in the first place.

No wonder campaigns underperform.

The Real Question

Here's what all of this comes down to:

When you ask "What's the industry benchmark?", you're actually asking this: "How did campaigns perform, on average, across a bunch of different companies with different audiences, different messages, different contexts, different goals, measured in different ways, by people with different definitions of success?"

That question has an answer. But the answer doesn't help you.

The question you actually need answered is this: "What happened in situations genuinely comparable to mine?"

Not "B2B SaaS campaigns"—that's a category, not a comparison.

Not "campaigns to CFOs"—that's still too broad.

Not even "campaigns to mid-market CFOs in financial services"—you're getting closer, but you're still missing most of what matters.

You need: "Campaigns with similar audience maturity, similar competitive intensity, similar message sophistication, similar company stage, similar engagement history, similar buying context, similar channel execution, and similar business goals."

Situations where the dimensions that matter actually *rhyme*.

Not campaigns that rhyme—most campaigns are too complex, too multifaceted to rhyme as complete entities. You need to look at campaigns that share rhyming dimensions across the variables that actually drive performance.

That's not a benchmark. That's not a category. That's not an average.

That's pattern recognition across dimensions that matter.

And it requires a fundamentally different approach—one that the industry isn't set up to provide, because it would mean admitting that everything they've been selling you is statistical theater.

So they keep publishing benchmarks. And marketers keep optimizing against them. And campaigns keep underperforming. And everyone blames execution instead of questioning the measurement.

In the next chapter, we'll explore why smart people keep using broken tools—why benchmarks provide psychological comfort even when they provide zero practical value.

But first, sit with this uncomfortable truth:

Every time you compared your campaign to an industry benchmark, you were comparing yourself to a number that doesn't describe any real campaign, computed from a sample that doesn't represent your situation, using an average that hides all the variance that matters.

You were measuring against fiction.

And calling it data-driven marketing.

You weren't lying to your clients or your manager when you reported those numbers. The benchmarks gave you permission to believe you were doing the right thing. You were measuring what everyone measures. You were comparing what everyone compares.

You were just lying to yourself about what the numbers meant.

And that's the most expensive lie of all.

Chapter Three

The False Comfort of Benchmarks

THE QUARTERLY BUSINESS REVIEW started exactly on time.

Kara Jensen, VP of marketing at a Series C enterprise software company, advanced to her slide deck. The dashboard glowed with achievement: email open rates up 7% quarter-over-quarter, click-through rates up 4%, lead generation costs down 12%. Every metric tracking above the horizontal benchmark lines.

"We're consistently outperforming industry standards across all channels," she said, voice steady with the confidence that comes from green dashboards and favorable variance reports. "Our demand generation campaigns delivered 2,847 MQLs this quarter, exceeding our target by 18%."

The CFO glanced up from his laptop. "And these benchmarks—where do they come from?"

"Industry standard metrics," Kara replied. "B2B SaaS aggregated performance data. We're comparing ourselves against companies at similar stage and scale."

He nodded and returned to his spreadsheet. The room relaxed slightly.

"The revenue analysis will be presented after lunch," Kara continued, clicking to the final slide. "Marketing's contribution to pipeline is tracking according to plan."

Nobody asked about the conversion rates that sales had been complaining about. Nobody mentioned that the competitive win rate had dropped 22% year-over-year, or that the company's most aggressive competitor had just stolen three major deals—all from accounts that had engaged with Kara's marketing campaigns.

Those questions would come up in the afternoon session. But they'd be framed as sales execution issues. Maybe the BDRs weren't following up fast enough. Maybe the sales process needed refinement. Maybe they needed better enablement materials.

Marketing had hit its numbers. The benchmarks proved it. Whatever revenue concerns emerged after lunch would be sales' problem to solve.

The QBR adjourned for the morning. Everyone left satisfied.

Nobody got fired.

And that's the entire point.

The Illusion of Certainty

Marketing is a profession built on uncertainty.

You launch a campaign in September targeting mid-market CFOs in financial services. Opens look good. Clicks seem reasonable. Three months later, sales reports that none of those leads converted. Was it the message? The timing? The audience targeting? The nurture sequence? The competitive landscape? The economic environment? The sales follow-up approach? All of the above? None of the above?

You'll never know for certain.

You can run attribution models, but they're approximations based on assumptions about how buyers actually make decisions—assumptions that are frequently wrong. You can A/B test, but you're testing variations within a strategic framework that might itself be flawed. You can ask prospects why they didn't buy, but they'll tell you the socially acceptable story rather than the uncomfortable truth.

Marketing decisions happen in a fog of incomplete information, delayed feedback, and confounded variables. You make your best judgment, execute, wait months for results, and then argue about what the results mean.

It's uncomfortable. Deeply, persistently uncomfortable.

The human brain despises uncertainty. We're pattern-seeking machines built by evolution to make quick decisions in environments where hesitation meant death. "Is that movement in the grass a predator or just wind?" You can't wait three months for more data—you need an answer *now*.

THE FALSE COMFORT OF BENCHMARKS

So we create certainty where none exists.

Benchmarks feel like certainty. They're numbers. They're based on "data." They're published by reputable sources. They come with charts and confidence intervals and year-over-quarter trend lines. They look scientific. They sound authoritative. They make the uncertainty tolerable.

"Our open rate is 24%, and the industry benchmark is 21%." There. You've transformed an ambiguous outcome into a clear win. You're not guessing—you're measuring. You're not hoping—you're validating. You're not operating in darkness—you have a reference point.

Except, as we established in Chapter 2, that reference point describes nothing real. It's a statistical artifact averaging together thousands of incomparable situations. It provides the *feeling* of certainty without providing actual insight.

But here's the uncomfortable truth: sometimes the feeling of certainty is more valuable than actual insight.

When you're presenting quarterly results to a board of directors who don't understand the nuances of marketing attribution, "We beat the benchmark" is a much better answer than "I believe our strategy is working, but the measurement systems are too noisy to prove it definitively."

When you're defending your budget allocation to a CFO who views marketing as a cost center, external validation beats internal conviction every time.

When you're trying to demonstrate professional competence in your annual review, pointing to metrics that outperform industry standards is safer than explaining why the metrics everyone else uses are fundamentally flawed.

We know benchmarks are problematic. We know they're averaging incomparable things. We know they're measuring what's easy rather than what matters.

We use them anyway because the alternative—admitting we're operating with incomplete information and making judgment calls—feels professionally dangerous.

Better to be precisely wrong than approximately right with no evidence.

At least when you're precisely wrong, you've got a chart.

And when you've only got 4.3 years to prove your value—the average tenure of Fortune 500 CMOs, according to Spencer Stuart—you don't have time to build sophisticated measurement systems. You need quick wins. You need numbers that trend favorably quarter-over-quarter. You need external validation that can be presented to the board.

Benchmarks provide that. They provide it fast. They provide it consistently. They provide career safety in a profession where average tenure is measured in years, not decades.

The CYA Economy

Let's be honest about what happened in Kara's quarterly business review.

She wasn't primarily trying to demonstrate business impact. She was trying to position marketing as successful regardless of what happened in the afternoon revenue session.

And she succeeded.

Think about how this dynamic actually plays out in organizations:

> **Scenario 1**: Your campaign generates a 26% open rate. The benchmark is 22%. Sales converts 0.8% of those leads into revenue.
>
> Marketing wins. You beat the benchmark. The dashboard is green. When sales complains about lead quality in the afternoon session, you've got a defense: "Marketing exceeded performance targets. The issue appears to be downstream in the sales process."

> **Scenario 2**: Your campaign generates a 19% open rate. The benchmark is 22%. Sales converts 3.4% of those leads into revenue—the highest conversion rate in company history. Revenue crushes targets.
>
> Marketing also wins! Revenue success speaks for itself. Nobody even looks at the benchmark gap. "Sure, opens were slightly below industry average, but that's just noise. What matters is that we drove record revenue." The benchmark was useful when you beat it and irrelevant when you missed it but revenue came through anyway.

This is what marketers are actually optimizing for: positioning themselves as winners no matter what happens.

THE FALSE COMFORT OF BENCHMARKS

But there's an uncomfortable question almost nobody asks: *What did you actually learn?*

In Scenario 1, you beat the benchmark, with terrible business outcomes. What does that tell you to do differently next quarter? Optimize your subject lines more? Double down on the tactics that drove that high open rate? You've succeeded according to the measurement system, so the feedback loop says "do more of this."

In Scenario 2, you missed the benchmark, with exceptional business outcomes. What does that tell you to replicate? You can't point to the tactics that worked because, according to your measurement system, they didn't work. The benchmark says you underperformed, even though you actually crushed it.

Either way, you're flying blind. The measurement system isn't teaching you anything about what actually drives results. It's just providing political cover.

It's theater. Expensive, time-consuming, data-rich theater.

And when next quarter rolls around, will your performance be determined by strategic insight or statistical luck? You won't know. Because you're measuring against a number that has no causal relationship to business outcomes.

Think about it: there's a strong causal relationship between marketing execution and sales results. The quality of your leads, the resonance of your messaging, the sophistication of your targeting—these directly impact whether sales can convert opportunities. But benchmarks don't measure any of that. They measure surface-level engagement metrics that may or may not correlate with the things that actually matter.

It's like trying to predict marathon performance by measuring how enthusiastically runners tie their shoelaces. There might be some correlation—enthusiastic shoelace-tying might indicate general readiness—but it's not causal. And optimizing for shoelace-tying enthusiasm won't make you run faster.

But if everyone in the industry measured shoelace enthusiasm, and published benchmarks for it, and evaluated runners based on it, and built entire compensation structures around it... well, you'd have a lot of runners optimizing their shoelace-tying technique while their actual race performance stagnated.

Squirrel!

That's what benchmarks do. They divert attention to the shiny object—the number everyone recognizes, the metric everyone measures—while the actual causal factors driving business outcomes hide in plain sight, unmeasured and unoptimized.

The Unknown Ceiling

Here's something that keeps me awake at night: What if your campaign that "beat the benchmark" was actually terrible?

You hit 26% opens. The benchmark was 22%. Celebration time.

But what if campaigns with similar audience characteristics, similar competitive context, similar message sophistication, similar timing—campaigns whose dimensions actually rhyme with yours—were achieving 41% opens?

You'd never know.

The benchmark told you that you won. So you stopped pushing. Why experiment with radically different approaches when you're already above average? Why take creative risks when the data says you're succeeding?

You're on cruise control, steadily optimizing incremental improvements to maintain your position above the benchmark line. Meanwhile, companies that abandoned benchmarks three years ago are discovering what's actually possible with your exact audience in your exact context.

They're learning. You're maintaining.

They're finding the ceiling. You're celebrating being above the middle.

And the gap between you grows wider every quarter, but you won't notice it in your dashboards. Because your dashboards compare you to "average B2B SaaS performance," not to what's actually achievable with campaigns whose dimensions rhyme with yours.

This is the cruelest trick benchmarks play: they don't just mislead you about whether you're winning or losing. They prevent you from ever discovering your true potential.

You're comparing yourself to a fiction when you should be learning from campaigns that share your actual characteristics across dimensions that matter. You're measuring against everyone, which means you're learning from no one.

And every quarter you spend celebrating above-average performance is a quarter you could have spent discovering what exceptional looks like.

The benchmark isn't just wrong. It's the ceiling on your ambition.

The Streetlight Effect

There's an old joke about a drunk man searching for his keys under a streetlight. A police officer asks, "Is this where you lost them?" The drunk replies, "No, I lost them in the park, but the light is better here."

This is exactly what benchmarks do to marketing measurement.

We don't measure what matters. We measure what's easy to measure. And then we optimize for what we measure. And then we create benchmarks for what we optimized. And then we judge ourselves against those benchmarks.

Display ad CTRs have benchmarks because clicks are easy to track at scale. You can aggregate millions of data points, calculate averages, segment by industry, and publish charts. The technology infrastructure is built for it. The data is clean. The metrics are standardized.

Message resonance doesn't have a benchmark. Competitive positioning strength doesn't have a benchmark. Strategic alignment with business goals doesn't have a benchmark. Long-term brand perception shift doesn't have a benchmark.

Not because these things don't matter—they matter *more* than clicks. But because they're hard to measure consistently across thousands of campaigns. They're subjective. They're contextual. They're confounded by dozens of variables.

So we don't measure them. Or we measure them occasionally, through expensive surveys and focus groups, but we don't make them the primary metrics. They don't go in the dashboard. They don't get tracked quarter-over-quarter. They don't have benchmark comparisons.

And slowly, imperceptibly, what we measure becomes what we value.

The VP of marketing doesn't ask, "Did this campaign strengthen our competitive positioning with enterprise buyers?" She asks, "Did we hit our CTR target?"

The agency account director doesn't say, "Let's take a creative risk that could differentiate your brand." He says, "Let's A/B test display creative to lift clicks by 3%."

The marketing coordinator doesn't think, "Is this message going to resonate with our actual target buyers?" She thinks, "Will this pass through ad verification and hit our viewability benchmark?"

We've optimized an entire profession around the things that are easiest to measure, not the things that matter most.

And the tragedy is that the things we don't measure—strategic fit, message sophistication, competitive differentiation, long-term relationship building—are the things that actually drive business outcomes.

But they don't have benchmarks. So they've become invisible.

We're searching for our keys under the streetlight, fully aware we lost them in the park, but too afraid to venture into the darkness where the light doesn't reach.

At least under the streetlight, we can show we're searching.

The Insight Theater

Let's talk about campaign insights for a moment.

Most agencies deliver them in end-of-campaign reports. Beautiful decks, comprehensive analyses, carefully selected learnings that position the agency's strategic value while setting up the pitch for next quarter's campaign.

"We learned that subject lines with numbers perform 18% better." "We learned that Tuesday sends outperform Thursday sends." "We learned that our target audience responds better to aspirational messaging than practical messaging."

These get presented as insights. They go in the deck. They inform next quarter's strategy. Everyone nods thoughtfully.

But are these actually insights, or are they artifacts of measurement theater?

Because real insights happen every day during campaign execution. Every A/B test failure is an insight—it tells you something didn't work and forces you to understand why. Every unexpected performance spike is an insight. Every message that resonates differently than predicted is an insight.

But most of those daily insights never make it into the end-of-campaign report. Because they're messy. Because they contradict the narrative. Because they don't fit neatly into the "what we learned" slide that justifies next quarter's budget.

The end-of-campaign report is optimized for selling the next campaign, not for capturing actual learning.

And benchmark comparisons make this worse. When your primary "insight" is "We beat the benchmark by X%," you're not learning anything about what actually drives performance in your specific context. You're just confirming that you successfully gamed the metrics everyone else is gaming.

A real insight would look like this: "Campaigns targeting prospects in our specific competitive context with our level of brand recognition performing this particular message strategy saw conversion rates 3x higher than our current approach—here's what we should test next quarter to replicate that pattern."

But that requires pattern recognition across dimensions that matter. It requires comparing your campaign to situations that actually rhyme with yours, not to statistical averages of everyone doing vaguely similar things.

Most agencies can't deliver that insight because they don't have access to that pattern recognition. So they deliver what they can measure: benchmark comparisons and surface-level tactical learnings.

And everyone calls it insight because we don't have language for what actual insight would look like.

The Ecosystem That Won't Let Go

Let's zoom out and look at why this system is so hard to escape.

It's not just that individual marketers are making bad decisions. It's that the entire professional ecosystem is designed to perpetuate benchmark-driven thinking.

Marketing agencies don't write contracts around benchmark delivery—that would be too obvious. Instead, they sell themselves on creative excellence and strategic prowess. They win the business with compelling case studies and sophisticated positioning.

Then three months later, the first client dashboard review happens. And somehow, mysteriously, every slide includes "vs. benchmark" comparisons. Nobody asked for this. Nobody specified it in the statement of work. But there it is—the universal language of marketing performance evaluation.

Because agencies know that's how clients actually evaluate success. The creative brilliance and strategic sophistication that won the pitch get translated into benchmark performance in the ongoing relationship. It's just how the game is played.

Marketing technology platforms sell dashboards full of benchmark comparisons. "See how your performance stacks up against similar companies." The benchmark feature is a selling point. It's how they demonstrate the value of their data aggregation. Admitting that benchmarks are misleading would undercut their core value proposition.

Marketing education—business schools, bootcamps, certification programs—teaches benchmark-driven decision-making as "data-driven marketing." Students learn to measure against industry standards. It's in the curriculum. It's on the exam. It's how you demonstrate analytical thinking.

Professional advancement is tied to demonstrating quantifiable results. "Increased email performance 23% above industry benchmark" looks great on a resume. "Made strategic decisions that I believe improved business outcomes, but I can't prove this with standardized metrics" does not. The job market rewards people who speak the language of benchmarks.

Executive education for chief executive officers and board members emphasizes the importance of marketing accountability. "How do you know marketing is working?" The answer everyone learns is this: "Compare performance to industry benchmarks." It's presented as sophisticated governance.

And remember: you've only got 4.3 years on average to prove your value as a CMO at a Fortune 500 company. You don't have time to educate your board on why benchmarks are misleading and build alternative measurement frameworks and change how the entire organization thinks about marketing effectiveness.

You need quick wins. You need numbers that trend favorably. You need to speak the language your board understands.

So you use benchmarks. Because everyone else uses benchmarks. Because that's what leadership expects. Because challenging the system takes time you don't have.

You've got an entire industry—vendors, agencies, consultants, educators, recruiters, thought leaders—all economically incentivized to keep the benchmark machine running.

This situation could be framed as a collective action problem. Most people in the industry sense that benchmarks are flawed. If you talk to marketers privately, off the record, many will admit discomfort with benchmark-driven measurement. "Yeah, I know it's not perfect, but it's what leadership wants to see."

But nobody wants to be the first to stop using them.

The agency that tells clients, "We don't measure against benchmarks" loses deals to agencies that show up with "vs. benchmark" dashboards. The marketing director who tells her CMO, "I can't give you a benchmark comparison" gets replaced by someone who can. The platform that removes benchmark features loses customers to competitors who kept them.

So everyone keeps playing the game, even though many people sense the game is problematic.

We've created a market failure. Individually rational behavior (using benchmarks for career protection) produces collectively irrational outcomes (industry-wide optimization for meaningless metrics).

And the language itself has calcified around benchmarks. When a board member asks, "How's our marketing performing?", the expected answer involves benchmark comparisons. We don't have widely accepted alternative frameworks. We don't have a shared vocabulary for discussing campaign effectiveness without reference to industry averages.

Try having a conversation about marketing performance without using the words "benchmark," "industry average," "above target," or "compared to similar companies." It's shockingly difficult. The conceptual infrastructure doesn't exist.

So we keep using the words we have, which forces us to keep using the frameworks behind those words, which perpetuates the system we all sense is broken.

Breaking out requires not just individual courage. It requires building new language, new frameworks, new measurement approaches, new vendor relationships, new agency contracts, and new board reporting formats.

It requires changing how an entire industry thinks about marketing effectiveness.

And that's hard. Really, really hard.

Especially when you've only got four years to prove yourself.

So we don't do it. We keep optimizing email subject lines and celebrating when we hit 26% open rates.

The Courage to Be Wrong Alone

Here's what it actually takes to abandon benchmarks.

You have to walk into a board meeting and say, "I can't compare our performance to industry benchmarks because benchmarks aggregate incomparable campaigns into meaningless averages."

You have to tell your CEO, "I believe our strategy is working based on pattern recognition across campaigns whose dimensions rhyme with ours, but I can't give you a simple number that shows we're beating the competition."

You have to explain to your CFO, "The metrics we should be tracking are harder to measure and don't have external validation, but they're actually predictive of business outcomes."

You have to accept that some people in that room will think you're making excuses. That you're avoiding accountability. That you're complicating things that should be simple.

You have to be comfortable with ambiguity in a professional environment that demands certainty.

You have to trust your judgment enough to stand behind it even when you can't point to an external benchmark that validates your decisions.

This is not a small ask. This is a career-defining risk.

There's an asymmetry in terms of consequences: If you're right, the business outcomes improve—but slowly, over time, in ways that are hard to attribute specifically to your decision to abandon benchmarks. If you're wrong, you look like the person who rejected proven measurement approaches and failed.

The upside is diffuse. The downside is sharp and personal.

And yet, the alternative—continuing to optimize against meaningless benchmarks while competitors steal market share—is equally risky. It's just slower. More comfortable. Easier to defend in the moment.

Being truly data-driven means having the courage to reject bad data.

It means being willing to say, "This number looks authoritative, but it doesn't actually tell us anything useful."

It means building measurement systems that are harder to explain but more accurate in practice.

It means accepting that sophistication sometimes looks like ambiguity to people who don't understand the nuance.

Most marketing leaders aren't willing to take that risk. And honestly, given the organizational dynamics we've just explored—including that 4.3-year average tenure—it's hard to blame them. The system punishes people who challenge benchmarks, even when they're right.

But keep this in mind: your competitors are making the same calculation. Most of them are optimizing against the same benchmarks, gaming the same metrics, celebrating the same hollow victories.

The companies that break free—that build genuinely sophisticated measurement systems based on pattern recognition rather than generic averages—are quietly building insurmountable competitive advantages.

While you're optimizing for 26% open rates, they're identifying which message strategies actually drive purchase decisions. While you're celebrating above-benchmark click rates, they're recognizing patterns across campaign dimensions that reveal what actually works in specific competitive contexts.

They're getting smarter. You're getting more consistently mediocre.

And by the time you notice the gap, it's too late to catch up.

The Question We've Been Avoiding

So let's acknowledge the elephant in the room.

If benchmarks are statistically meaningless, psychologically comforting but strategically misleading, and structurally embedded in how the entire industry operates...

What the hell do you measure instead?

That's the question we've been circling for two chapters. And it's the question most marketers can't answer. Which is exactly why they keep using benchmarks. Because benchmarks are a bad answer, but at least they're *an* answer.

The honest truth is this: you can't just stop using benchmarks without replacing them with something better. You need an alternative framework. A different way of evaluating performance. A new language for discussing what success means.

You need to stop asking, "How do I compare to everyone?" and start asking, "What campaign dimensions genuinely rhyme with mine across variables that actually matter?"

You need to stop treating marketing as a categorization exercise—"We're B2B SaaS, therefore we should perform like other B2B SaaS companies"—and start treating it as a pattern recognition challenge.

Because here's what I've learned from 25+ years of B2B marketing: complete campaigns don't rhyme with complete campaigns. That's too simplistic. Too surface-level.

What rhymes are *dimensions* of campaigns.

Your audience maturity might rhyme with someone else's audience maturity, even though you're in different industries. Your competitive context might rhyme with their competitive context, even though you're selling different products. Your message sophistication might rhyme with their message sophistication, even though you're targeting different titles.

You need to find campaigns that share similar characteristics across the dimensions that drive performance—audience behavior patterns, competitive intensity levels, message strategy approaches, channel execution sophistication, timing dynamics, and market conditions.

Not campaigns in the same industry bucket. Not companies at a similar revenue scale. Dimensions that actually rhyme across the variables that matter.

That's not a benchmark. That's not an average. That's not a category.

That's multi-dimensional pattern recognition.

And it requires seeing marketing in a fundamentally different way—not as a set of tactics to be optimized against averages, but as a complex system where success patterns repeat across non-obvious dimensions.

Most of the industry isn't ready for that conversation. They're still searching for their keys under the streetlight. They're still celebrating wins that might actually be losses. They're still producing insights that are really just sales theater for the next campaign pitch.

But you're different. You've read this far. You've sat with the uncomfortable reality that everything you've been measuring might be fiction. You've acknowledged your own complicity in the system.

You're ready to ask better questions.

The Agency Advantage (That Nobody's Actually Using)

Before we move to Part 2, let's address something that's causing existential anxiety across the marketing industry right now: the great in-house versus agency debate.

With AI-powered marketing tools promising to democratize everything from content creation to campaign optimization, more brands are asking, "Why do we need an agency? Can't we just do this ourselves?"

It's a fair question. And the answer depends entirely on what "this" means.

If "this" means executing tactics—writing ad copy, building landing pages, running A/B tests—then yes, AI-powered SaaS tools have made in-house execution more feasible than ever. A talented marketing coordinator with the right software stack can produce a tremendous volume of campaign assets.

But if "this" means *learning from patterns across diverse campaigns to make your specific campaigns smarter*, then agencies have a structural advantage that in-house teams simply can't replicate.

And yes, I'm inherently biased here. I spent 25+ years in agency and partner marketing roles. I believe agencies provide unique value.

But I'm even more biased toward agencies that actually *deliver* that unique value to their clients. Which, spoiler alert, basically don't exist. Which is why I built a company to do exactly that.

Here's why agencies *should* have an advantage: a brand only works on a brand, but an agency is a house of brands.

When you bring marketing in-house, you gain control and alignment. You deeply understand your product, your customers, and your competitive context. You can move fast and iterate without external coordination overhead.

But you also develop blind spots that are hard to see from the inside.

Watch what happens when brands bring creative in-house: within six months, the company tagline starts appearing as the headline in every campaign. "Empowering Innovation Through Cloud Solutions" becomes the hero message on display ads, email subject lines, and landing pages.

But a tagline isn't a headline. A tagline is brand positioning—it describes who you are. A headline is campaign messaging—it describes why someone should care *right now* about *this specific thing* you're offering.

In-house teams, deeply immersed in the brand, start to conflate the two. They lose perspective on the difference between corporate identity and campaign provocation.

Agencies—when they're functioning properly—maintain that outside perspective. They see your brand as one of many they work with, which helps them recognize when you're defaulting to safe, generic positioning instead of sharp, specific messaging.

But there's a larger strategic advantage that agencies possess—and almost never leverage.

Your in-house team experiences one competitive landscape, one set of buyer behaviors, one messaging framework, and one set of channel dynamics. They get very, very good at optimizing *your* campaigns within *your* context.

An agency working across ten clients accumulates ten different competitive landscapes. Ten different sets of buyer behaviors. Ten different messaging frameworks. Ten different channel execution contexts.

More importantly, they accumulate multi-dimensional data—performance patterns across audience types, message strategies, competitive intensities, buying stages, channel combinations, timing dynamics, and dozens of other variables that create the signature of campaign performance.

But *they're not using it*.

Most agencies organize their teams by client. The people working on your brand work *only* on your brand. The team working on their healthcare client never talks to the team working on their financial services client. The insights from the HR software campaign never make it to the finance software campaign—even when the campaigns share rhyming dimensions around buyer psychology, message sophistication, and competitive positioning.

The multi-dimensional data exists. The pattern recognition opportunity exists. But the organizational structure, client confidentiality concerns, and sheer operational complexity prevent agencies from actually mining those patterns at scale.

And even if they could organize for it, most agencies lack the foundational AI infrastructure to extract dimensional patterns across campaigns, maintain strict privacy boundaries, and translate those patterns into actionable insights for specific clients.

So the structural advantage exists—agencies absolutely *should* be able to deliver pattern intelligence that in-house teams can't access. But they're not doing it. Either because they're not thinking dimensionally (they're still trapped in category-based thinking like "B2B SaaS" or "financial services"), or because they've organized their teams in ways that prevent cross-client pattern recognition, or because they lack the AI-native infrastructure to do this at scale while maintaining client confidentiality.

An in-house team might run 15 campaigns per year—all for the same product, same brand, same market. An agency might run 150 campaigns per year across 10 different clients in diverse industries. The pattern recognition possibilities aren't just 10x larger—they're exponentially larger because of the dimensional diversity.

But if that dimensional diversity sits locked in separate client silos, never aggregated or analyzed for patterns, the advantage remains theoretical.

And here's where we get to the real opportunity—both for agencies and for brands working directly with partners who can provide this intelligence.

Imagine an agency—or an AI-native marketing intelligence partner—that *actually* leverages that multi-dimensional data. That recognizes when your message sophistication level combined with your audience's buying stage and your competitive positioning creates a dimensional signature that rhymes with three other campaigns they've run—maybe none in your industry—and here's what those campaigns learned that you should test.

That's not something an AI tool can deliver if it's only analyzing one brand's data. That's not something an in-house team can discover working solely within their own context. And that's not something traditional agencies are structured to provide, even though they theoretically possess the raw material.

And here's something crucial to understand about AI in this context: off-the-shelf AI tools aren't built for dimensional pattern recognition across marketing campaigns.

Sure, you can feed 20 campaign performance spreadsheets into ChatGPT or Claude and ask, "What patterns do you see?" The AI will dutifully analyze the data, identify statistical correlations, and generate insights.

But those insights are only as sophisticated as the questions you know to ask—and the frameworks you've built through experience in recognizing what actually matters.

Because here's the uncomfortable reality: people—and machines—can make any data tell whatever story they want to tell. Doesn't mean it's a true story.

You can cherry-pick time periods, emphasize convenient metrics, ignore inconvenient context, and arrive at conclusions that look data-driven but are actually narrative-driven. AI is exceptionally good at finding patterns in data. It's also exceptionally good at finding patterns that don't actually exist—correlations that are statistical artifacts rather than causal relationships.

Without experienced human judgment guiding what to look for and how to interpret what's found, you get sophisticated-sounding insights that are just expensive fiction.

AI doesn't inherently know that message sophistication level matters more than industry category. It doesn't know which dimensional combinations create meaningful rhymes and which are statistical noise. It doesn't know that a campaign targeting risk-averse compliance buyers in healthcare might share critical success patterns with a campaign targeting risk-averse compliance buyers in financial services—even though "healthcare" and "financial services" benchmarks would treat them as completely different.

That knowledge comes from human expertise. In my case, it comes from 25+ years of watching campaigns succeed and fail. From recognizing patterns that aren't obvious in the spreadsheet but become clear when you've lived through countless similar situations. From understanding buyer psychology, competitive dynamics, and organizational decision-making processes in ways that no training data can capture.

The AI is extraordinarily powerful at *scale*—at processing volumes of campaign data that no human could analyze manually, at calculating similarity scores across dozens of dimensions simultaneously, at generating creative variations once you've identified the strategic pattern.

But the *instruction set*—knowing which dimensions to measure, how to weight them, which patterns actually predict success, how to translate statistical similarity into strategic insight—that requires human judgment refined through extensive real-world experience.

This is why pattern intelligence at scale requires both: AI to process the dimensional complexity that humans can't handle manually, and human expertise to guide what the AI should look for and how to interpret what it finds.

The in-house team lacks dimensional diversity. The traditional agency has dimensional diversity but isn't structured to leverage it. And off-the-shelf AI tools lack the domain expertise to know what patterns matter.

You need all three: dimensional diversity, AI infrastructure, and deep marketing expertise that guides the analysis.

That's pattern intelligence at scale. That's what dimensional rhyming actually looks like in practice.

For in-house teams, this means partnering with someone who *can* provide that cross-campaign pattern intelligence—supplementing your deep brand knowledge with insights from dimensional patterns across diverse contexts.

For agencies, this means either radically restructuring how you organize, share insights, and leverage AI infrastructure... or partnering with specialists who've built their entire operation around dimensional pattern recognition from day one.

So yes, I'm biased toward the value that agencies *should* provide. But I'm insanely biased toward agencies that actually *deliver* this level of intelligence to their clients.

Which don't really exist.

Which is why I built a company designed to do exactly that—for agencies who want to add this capability, and for brands who need this intelligence regardless of whether they work with agencies or in-house teams.

Because the advantage of multi-dimensional pattern intelligence is too valuable to remain theoretical.

Someone needs to actually make it real.

Which brings us back to the central question: how do you actually do that?

That's what Part 2 is for.

Bridge to the Alternative

We've established the problem completely:

- **Chapter 1**: Benchmarks create expensive illusions of success (emotional stakes).

- **Chapter 2**: Benchmarks fail mathematically by averaging incomparable campaigns (intellectual foundation).

- **Chapter 3**: We use benchmarks anyway because they provide psychological safety, political cover, and career protection—despite preventing us from ever discovering our true potential (systemic analysis).

The groundwork is complete. You understand *why* industry averages destroy campaign performance. You understand *how* aggregation fails. And you understand *why* smart people keep using broken tools even when they sense something's wrong.

Now we can introduce the alternative.

Part 2 reveals what actually works: finding campaign dimensions that rhyme across variables that matter, building intelligence from genuinely comparable situations, and measuring performance against patterns that predict success rather than averages that describe nothing.

This is where the uncomfortable truth becomes useful.

Let's begin.

PART 2
THE ALTERNATIVE

Chapter Four

Multi-Dimensional Pattern Recognition

As you may recall from the Author's Note at the beginning of the book, there was a moment that changed everything for me.

I was reviewing performance data from two completely unrelated campaigns—one targeting chief compliance officers at mid-market insurance companies, the other targeting IT decision-makers at large retailers. Different industries. Different titles. Different products. Different team members working on them. Nothing obviously comparable.

Except both campaigns dramatically outperformed expectations in exactly the same way.

Not just "both did well." They *rhymed*. Same signature of engagement patterns. Similar curves in how prospects moved through nurture sequences. Parallel behaviors in how they consumed content before converting. Even the timing of their conversion triggers followed remarkably similar arcs.

The campaigns themselves didn't rhyme—one was selling regulatory compliance software through thought leadership content; the other was selling application performance monitoring through return on investment (ROI) calculators. Completely different value propositions, completely different creative approaches, completely different channels.

But their *dimensions* rhymed.

Both targeted risk-averse buyers in highly regulated environments who needed to build internal consensus before making purchasing decisions. Both faced intense competitive pressure from entrenched incumbents with deeper pockets. Both were selling complex solutions requiring behavioral change, not just technology adoption. Both audiences were sophisticated enough to smell generic marketing from a mile away.

And most critically, both campaigns succeeded by doing something their competitors weren't doing: acknowledging uncertainty rather than promising certainty.

The insurance compliance campaign opened with "You can't predict every regulatory risk—but you can prepare for the ones that matter most." The application performance monitoring campaign led with "No monitoring platform eliminates all incidents—but the right one eliminates the incidents that actually cost you money."

Different words. Same fundamental strategy. Same dimensional profile. Same success pattern.

No benchmark would have connected those dots. "Insurance" and "retail" live in completely different categories. "CCO" and "ITDM" are different buyer personas. The campaigns would never appear in the same industry report, never get averaged into the same benchmark, and never be considered comparable by traditional analysis.

But they were teaching the same lesson.

And once I saw it, I couldn't unsee it.

The Fundamental Shift

For decades, marketing intelligence has operated on a categorization paradigm: group campaigns by industry, by channel, by audience demographics, and by product type. Then calculate averages within those categories. Then compare your performance to the averages.

It's clean. It's intuitive. It's how humans naturally organize information.

And it's fundamentally wrong for marketing intelligence.

Because campaign performance doesn't organize neatly into categories. It emerges from complex interactions across *dimensions*—multiple variables influencing each other simultaneously in ways that simple grouping can't capture.

Think about how you actually describe a campaign that worked. You don't say "it was a B2B SaaS campaign." That tells you almost nothing. You say something like this:

"We targeted mid-market buyers who were already aware they had a problem but didn't trust that solutions actually worked. We led with customer skepticism testimonials—not success stories, but 'here's what we thought wouldn't work and were wrong about' content. We sequenced it so prospects got the skepticism content first, then the solution mechanics, then the ROI proof. And we used our sales team's authentic voice instead of corporate marketing speak because this audience had been burned by overpromising vendors before."

That description contains at least eight dimensions:

- Company size (mid-market)

- Problem awareness level (already aware)

- Trust position (skeptical of solutions)

- Content strategy (skepticism testimonials)

- Sequencing approach (skepticism → mechanics → ROI)

- Voice/tone (authentic vs. corporate)

- Competitive context (overpromising category)

- Audience psychology (burned before)

Those dimensions interact. The skepticism testimonials work *because* the audience has been burned before. The sales team voice works *because* prospects distrust corporate marketing. The sequence works *because* you're matching the audience's natural progression from skepticism to consideration.

Change any one dimension and the pattern might break. Remove the prior-burned context and skepticism testimonials might just seem negative. Use a corporate voice and the authenticity advantage disappears. Lead with ROI proof before addressing skepticism and prospects bounce immediately.

Now here's what makes this powerful: somewhere out there, another campaign is succeeding with a remarkably similar dimensional profile—but in a completely different industry, selling a completely different product, to a completely different job title.

And the lessons from that campaign can inform yours in ways that a benchmark never could.

Because dimensions rhyme even when campaigns don't.

Why Multi-Dimensional Similarity Works

Let's get concrete about why this matters.

Traditional benchmarks collapse dozens of meaningful dimensions into a few crude categories. "B2B SaaS to CFOs in financial services" captures maybe 5–7 dimensions at most. Everything else—message sophistication, competitive intensity, buying stage, risk tolerance, organizational complexity, prior vendor relationships, urgency level, budget authority—gets ignored.

But those ignored dimensions often matter *more* than the categories you used.

A campaign targeting skeptical, risk-averse CFOs in financial services might have more in common with a campaign targeting skeptical, risk-averse chief information security officers in healthcare than it does with a campaign targeting aggressive, growth-focused CFOs in financial services.

Same category, different dimensions, different patterns of success.

Multi-dimensional pattern recognition flips the logic: instead of forcing campaigns into pre-defined categories and averaging performance within those categories, you measure similarity *across* dimensions and identify campaigns whose dimensional profiles rhyme—regardless of what categories they fall into.

It's the difference between organizing books by spine color versus organizing them by themes that actually matter to readers. Color-coding is fast and looks neat on the shelf, but it tells you nothing about whether you'll enjoy a book. Theme-based organization requires more sophisticated analysis, but it actually helps you find what you're looking for.

Let me describe the mathematical intuition: when you average campaigns within a category, you're assuming that the category captures most of what matters about campaign similarity. But if the category captures only 15% of meaningful variance—and the other 85% comes from dimensions you're not measuring—then your average is mostly noise.

Pattern recognition across dimensions inverts that logic. You measure similarity across the variables that *actually predict* similar performance patterns. You're not asking, "What category does this campaign belong to?" You're asking, "What combination of dimensional characteristics does this campaign share with others that succeeded or failed?"

This is why pattern recognition finds insights that benchmarks miss. You're not constrained by predetermined categories. You're discovering *emergent* similarities that only become visible when you analyze multiple dimensions simultaneously.

The aspect of this approach that makes this even more challenging is also what makes it so powerful in practice: the dimensions that matter most aren't stable—they shift depending on what question you're trying to answer.

If you're optimizing email subject lines, message strategy dimensions and prior engagement patterns matter enormously. Company size? Barely relevant. Industry? Mostly noise.

If you're optimizing channel mix, business goals and sales cycle length dominate. Message sophistication? Less critical at that level.

If you're optimizing targeting, competitive intensity and buyer awareness stage matter far more than the specific product category.

Pattern recognition doesn't force you to pick which dimensions matter in advance. It lets you weight dimensions contextually based on what you're trying to learn.

The Rhyming Metaphor

I've been using this phrase repeatedly: "campaigns that rhyme." Let me be precise about what I mean.

In poetry, rhyming isn't about identical words—it's about similar phonetic patterns. "Light" and "fight" rhyme not because they're the same word or even related meanings, but because they share acoustic structure.

There's a brilliant moment from a 2010 *60 Minutes* interview with Eminem where Anderson Cooper asks him about "bending words." Eminem's response perfectly captures what I'm talking about. Addressing a common claim, he says: "People say that the word orange doesn't rhyme with anything... and that pisses me off because I can think of a lot of things that rhyme with orange." He then demonstrates how you can bend syllables and manipulate phonetics to make "orange" rhyme with "door hinge," "four inch," "storage"—showing that rhyming is about creative pattern recognition, not rigid dictionary rules.

That interview sparked news articles, Reddit debates, blog posts, video reactions—from people across demographics. Hip hop fans and classical music listeners. People who'd never heard of Eminem and longtime followers. Why such broad engagement? Because nobody thought about rhyming that way. Everyone accepted the conventional wisdom that orange was unrhymable. Eminem saw patterns others missed by thinking dimensionally about sound rather than categorically about spelling.

You can watch the clip here: https://go.cadenceb2b.com/eminem

Campaign patterns work the same way.

When I say campaigns rhyme, I mean they share dimensional signatures—similar profiles across variables that influence success—even when their surface characteristics look completely different.

That insurance compliance campaign and the application performance monitoring campaign? They didn't look alike. Different formats, different channels, different creative approaches, different language, different industries.

But their dimensions rhymed:

- Both targeted risk-averse consensus buyers

- Both faced entrenched competitive pressure

- Both required behavioral change, not just technology adoption

- Both succeeded by acknowledging uncertainty instead of promising perfection

- Both used an authentic voice to cut through corporate marketing noise

- Both sequenced content to match a natural skepticism-to-consideration progression

Those dimensional rhymes predicted that similar strategic approaches would work in both contexts—despite the surface differences.

This is the critical insight that unlocks pattern recognition: you're not looking for campaigns that *repeat*. Marketing rarely repeats. Audiences evolve; competitive contexts shift; channels change effectiveness; what worked last quarter might fail this quarter.

But campaigns *rhyme* all the time. Different industries, different products, different audiences—but similar dimensional signatures that predict similar success patterns.

The mistake most marketers make is looking for identical situations. "Show me another B2B SaaS campaign that targeted mid-market CFOs with email campaigns about cloud migration." That's searching for repetition, and you'll rarely find it.

The smarter question: "What campaigns succeeded with similar audience skepticism levels, similar competitive intensity, similar message sophistication requirements, similar consensus-building challenges—regardless of industry or product?" That's searching for rhymes.

And rhymes are everywhere once you know what to listen for.

Finding Actionable Insights Benchmarks Can't

Let's return to Sarah's $50K mistake from Chapter 1. Remember the problem: 847 leads from financial services executives, only 11 conversions, and everyone blamed everyone else.

The agency's solution? Segment by sub-industry. Retail banking gets different messaging than wealth management, insurance gets its own track.

That's still category thinking. You're just creating smaller categories to average within.

Here's what pattern recognition would reveal instead:

Those 11 successful conversions? They weren't random. When you analyze them dimensionally, they share a profile: all were relatively new in their roles (less than 18 months), all worked at companies that had recently experienced regulatory penalties or near-misses, all engaged most heavily with risk mitigation content rather than efficiency content, and all requested conversations within 72 hours of receiving content about competitive case studies in their specific sub-industry.

The 836 non-conversions? Mixed dimensional profiles. Some were long-tenured executives unlikely to champion new solutions. Some worked at companies without recent regulatory pressure. Some engaged with efficiency messaging that didn't match their actual pain. Some were served generic financial services content when they needed sub-industry specificity.

There's an insight to be gleaned here: if you look across *all* campaigns you've run—not just financial services, not just email, but everything—you'll find that new-in-role buyers dealing with recent crisis events who engage with risk mitigation content convert at 4–6x the rate of other prospects.

That pattern *rhymes* across industries, products, and channels. It's not specific to financial services. It's a dimensional signature.

And once you recognize it, you can do the following:

Prioritize ruthlessly

Stop treating all 847 leads equally. Route the ones matching that dimensional profile to senior BDRs immediately. Let automation handle the rest until they develop crisis urgency.

Test across dimensions

That risk-mitigation-content pattern probably works for manufacturing companies facing safety incidents, healthcare companies facing compliance issues, and retail companies facing data breaches. Test it.

Refine the messaging strategy

Lead with crisis relevance, not feature efficiency. Competitive case studies work when they match the prospect's specific context. Generic financial services content fails.

Optimize media allocation
Double down on channels and targeting approaches that reach new-in-role buyers at companies with recent regulatory events. Scale back on broad financial services targeting.

None of that comes from a benchmark. "B2B SaaS email campaigns to financial services" averages away the exact dimensional patterns that predict success.

But pattern recognition *across* dimensions reveals them clearly—and makes them immediately actionable.

Here's another example of a pattern hiding in plain sight:

An agency ran 15 different campaigns across five clients over 18 months. Traditional analysis by client showed mixed results—some campaigns worked, others didn't, with no clear pattern within any single client's campaigns.

The agency had rich contextual information about each campaign. If they'd been disciplined about it, they would have documented audience context, competitive dynamics, message strategies, and execution approaches in proper creative briefs—the kind of dimensional thinking smart agencies have always done at a basic level.

But those briefs were used only for creative development. Nobody thought to analyze *across* campaigns to find dimensional patterns. The multi-client data sat in separate folders, organized by client, never aggregated for pattern recognition.

What they missed: campaigns that used "decision framework" content (tools helping buyers evaluate options systematically) dramatically outperformed the others—but *only* when targeting buyers in highly competitive categories where multiple vendors looked similar.

That insight was invisible when analyzing by client or industry. It would have only emerged from looking across audience type, competitive context, and content strategy dimensions simultaneously.

And once identified, it would have been immediately actionable: if you're in a crowded category where differentiation is hard, lead with decision frameworks. If you're in a sparse category with clear differentiation, decision frameworks are less effective—lead with the unique value proposition instead.

That's not intuitive. It's not something you'd discover from benchmarks or best practice guides. But it's exactly the kind of pattern that emerges when you analyze dimensionally across diverse campaigns.

The agency had everything they needed to discover this. Multiple clients. Diverse campaigns. Rich contextual information (if documented in proper briefs). They just weren't thinking about pattern recognition across dimensions.

This is the opportunity most agencies are missing: they have multi-dimensional data from diverse client experiences, but they're organized in ways that prevent cross-campaign pattern recognition. And even when agencies feed campaign data into AI tools for analysis, they're typically only providing raw performance metrics—not the rich contextual information about audience, competitive environment, message strategy, and execution approach that makes dimensional analysis possible.

This is what makes dimensional pattern recognition so powerful: it surfaces *non-obvious* insights that work precisely because your competitors aren't looking for them.

The Five Core Dimensions (Preview)

I've been talking abstractly about "dimensions" throughout this chapter. In Chapter 6, we'll explore the complete dimensional framework in detail—how to map campaigns, how to calculate similarity, and how to weight different dimensions for different questions.

But here's the preview: after analyzing campaigns across 25+ years, I've found five core dimension categories that consistently predict performance patterns:

Dimension 1: Audience behavior

Who you're targeting, but measured in ways that actually matter—not just job title and industry, but awareness stage, risk tolerance, decision-making authority, competitive product experience, organizational dynamics, and buying urgency.

Dimension 2: Competitive environment

The market dynamics your campaign operates within—competitive intensity, product differentiation, category maturity, switching costs, incumbent strength, and budget availability.

Dimension 3: Message strategy

How you position your solution—core promise type, positioning approach, proof strategy, objection handling, emotional vs. rational emphasis, and sophistication level relative to audience.

Dimension 4: Execution approach

How you actually run the campaign—channel selection rationale, creative complexity, personalization depth, multi-touch sequencing, follow-up cadence, and measurement sophistication.

Dimension 5: Business goals

What you're actually trying to accomplish—primary key performance indicator (KPI) focus, sales cycle length, average deal size, conversion stage targeted, and success timeline.

Each dimension contains multiple sub-dimensions. Each sub-dimension can be measured. And most importantly, they interact—the right message strategy depends on audience context, the execution approach depends on the competitive environment, and business goals shape how you weight other dimensions.

When you map two campaigns across these five dimensions and calculate similarity scores, you can identify which campaigns genuinely rhyme—and therefore which lessons transfer.

A campaign with 87% dimensional similarity across audience context, competitive environment, and message strategy will teach you far more than a campaign in the same industry with 23% dimensional similarity.

That's not a benchmark. That's not a category. That's pattern recognition.

And it requires a fundamentally different analytical infrastructure—one that can measure dozens of dimensions simultaneously, calculate multi-dimensional similarity at scale, and surface rhyming patterns that humans couldn't spot manually.

Which brings us to the technical reality: this approach wasn't practically possible five years ago.

How AI Makes This Real

Humans have always performed versions of dimensional pattern recognition, intuitively.

When an experienced marketer says, "This reminds me of a campaign we ran three years ago," they're recognizing dimensional similarity, even if they can't articulate exactly which dimensions are rhyming. When an agency strategist says, "I've seen this pattern before in a completely different industry," that's pattern recognition across dimensions.

The problem is scale and consistency.

A human can maybe track patterns across 15–20 campaigns they've personally worked on. They can hold maybe 3–5 dimensions in working memory simultaneously. They can recall examples that share obvious similarities.

But they can't analyze 200 campaigns across 60+ dimensions simultaneously. They can't calculate similarity scores accounting for dimensional weighting and interaction effects. They can't surface non-obvious rhymes that only become visible in multi-dimensional space.

That's where AI—specifically, modern large language models and machine learning architectures—creates a breakthrough.

And I want to be precise here about what changed and when, because it matters for understanding why this approach is suddenly practical at scale when it wasn't five years ago.

The first wave of AI in marketing—let's call it the 2015–2020 period—focused on automation and efficiency. Programmatic ad buying. Automated email sends. Chatbots. Basic personalization. These tools made existing processes faster and cheaper, but they didn't fundamentally change *what* was possible analytically.

The second wave—the LLM revolution that accelerated dramatically in 2022–2023 and continues today—is different in kind, not just degree.

These models can ingest unstructured campaign descriptions, extract dimensional characteristics, map them into multi-dimensional vector spaces, calculate similarity across those spaces, and surface insights that would take humans months to identify manually.

More importantly, they can do this while maintaining context about *why* dimensions matter, what interactions exist between dimensions, and how to translate statistical similarity into strategic insight.

Here's a concrete example of what's now possible:

Modern AI systems can ingest campaign descriptions—not just performance data, but rich contextual information about audience, message strategy, competitive environment, execution approach, and business goals. They can map campaigns into multi-dimensional spaces where similar campaigns cluster together based on their dimensional signatures, then surface insights about which patterns matter most for specific strategic questions.

This type of intelligence wasn't practically achievable five years ago. The analytical infrastructure simply didn't exist to handle this level of complexity at scale while maintaining the contextual nuance that makes the insights actionable.

But here's what makes this breakthrough real rather than just theoretically possible: it requires sophisticated frameworks for dimensional analysis, not off-the-shelf AI tools. You can't simply feed campaign spreadsheets into ChatGPT and get useful dimensional pattern recognition. The infrastructure needs to be purpose-built for this specific analytical challenge.

And—this is absolutely critical—AI doesn't do this inherently. It requires several things working together:

First: You need to know *which* dimensions to measure. AI can't tell you that buying stage awareness matters more than job title unless someone with deep marketing expertise has built that knowledge into the analytical framework.

Second: You need to understand *how* dimensions interact. AI can calculate correlations, but it doesn't inherently know that message sophistication should be matched to audience sophistication, or that competitive intensity changes how you should weight execution complexity.

Third: You need to guide the AI on what patterns are *meaningful* versus what's statistical noise. AI will find patterns everywhere—some real, many spurious. Experienced human judgment is essential for knowing which patterns predict success versus which are coincidental.

Fourth: You need to translate statistical similarity into strategic action. "These campaigns have 84% dimensional similarity" is a data point. "Therefore, you should test leading with risk mitigation instead of efficiency benefits, and sequence your content to match natural skepticism-to-consideration progression" is strategic intelligence. That translation requires human expertise.

This is why I keep emphasizing that dimensional pattern recognition at scale requires *both* AI and experienced humans working in concert. The AI handles the complexity humans can't—simultaneous multi-dimensional analysis at scale across hundreds of campaigns. The humans handle the judgment AI can't—knowing which dimensions matter, how to weight them contextually, what patterns are meaningful, and how to turn patterns into strategy.

It's a genuine partnership. Not "AI does the work and humans just review it." Not "humans do the strategy, and AI just executes." Both are essential. Both bring capabilities that the other lacks.

And here's what makes this particularly exciting right now: AI capabilities in this domain are advancing *rapidly*. Every 6–12 months, the models get better at handling nuance, maintaining context, and surfacing non-obvious patterns.

Three years ago, you needed extensive prompt engineering and careful result curation to get useful dimensional analysis from AI. Today, the models are sophisticated enough to engage in genuine collaborative analysis—you can have a back-and-forth dialogue with the system, refining hypotheses, testing dimensional weightings, and exploring alternative interpretations.

Five years from now? The analytical capabilities will be even more powerful. More dimensions will be measurable. More subtle interactions will be detectable. More sophisticated pattern recognition will emerge.

But—and this is the strategic insight that matters—as AI capabilities advance, they become more accessible to everyone. The tools commoditize. What doesn't commoditize is the human expertise guiding what to look for and how to interpret what's found.

The Correlation Trap: Why AI Needs Human Judgment

Here's something crucial to understand about AI and pattern recognition: AI is exceptionally good at finding correlations. It will discover statistical relationships everywhere in your data—some meaningful, many completely spurious.

A classic example: ice cream sales and drowning deaths are highly correlated. AI analyzing that data would flag it as a strong pattern. But it's not causal—both are driven by a third variable (summer weather). Acting on that correlation would be absurd.

Marketing data is full of these traps. AI might notice that campaigns sent on Tuesdays at 10 am perform 15% better than campaigns sent on Thursdays at 2 pm. Looks like a pattern. Might even be statistically significant. But is it causal, or is it correlation driven by confounding variables you're not measuring?

Maybe Tuesday campaigns consistently target different audience segments. Maybe the Thursday campaigns had weaker subject lines. Maybe competitive activity spikes on Thursdays. Maybe your Tuesday data happened to avoid weeks with Monday holidays, when engagement is always lower, artificially inflating Tuesday performance. Maybe it's a pure coincidence across a small sample.

AI can't inherently distinguish correlation from causation. It sees patterns. It calculates statistical relationships. But it doesn't understand *why* patterns exist or whether they represent genuine causal mechanisms that will hold under different conditions.

This is where experienced human judgment becomes essential. After 25+ years of watching campaigns succeed and fail, I've developed intuition about which patterns represent real causal relationships versus which are statistical noise:

"That pattern makes sense because it aligns with known buyer psychology around decision-making under uncertainty."

"That correlation is probably spurious—it's only appearing in our data because of how we happened to structure our test, not because it represents a genuine insight."

"That dimensional similarity looks significant mathematically, but contextually those campaigns operated in fundamentally different competitive environments that would make the lessons non-transferable."

AI provides the computational power to surface patterns humans couldn't find manually. Human expertise provides the judgment to distinguish meaningful patterns from statistical coincidence.

Together, they enable dimensional pattern recognition that's both scalable and reliable. Separate them, and you either have limited human analysis that can't process multi-dimensional complexity, or sophisticated AI generating expensive nonsense.

We're not searching for correlations. We're searching for *causal* patterns—dimensional similarities that predict comparable performance because they represent genuine underlying mechanisms, not statistical artifacts.

That distinction is why this approach requires purpose-built infrastructure *and* deep marketing expertise working in concert. The infrastructure finds the patterns. The expertise validates which patterns matter.

This is the counterintuitive truth about AI in marketing intelligence: as AI gets better, experienced human judgment becomes *more* valuable, not less. Because everyone has access to similar AI tools, but not everyone has years of pattern recognition experience to know what questions to ask.

We'll explore this dynamic in much more depth in Chapter 7, where we'll examine why the human-AI partnership creates a strategic moat that appreciates over time rather than depreciating with each new model release.

For now, the key point is this: dimensional pattern recognition is practically possible at scale because of AI breakthroughs in the last 2–3 years. It will become even more powerful as AI continues advancing. But it requires experienced human expertise to be useful rather than just sophisticated-sounding nonsense.

You can't buy this capability as a commodity AI tool. You can't replicate it by just feeding your campaign data into ChatGPT. It requires purpose-built infrastructure, sophisticated frameworks for dimensional analysis, and deep marketing expertise guiding the entire process.

Which is exactly what makes it valuable.

A Brief Technical Preview

I'm not going to dive deep into the technical architecture here. But since we've been talking about AI enabling pattern recognition, it's worth previewing how this actually works under the hood.

The core technical approach uses *embedding models* to represent campaigns in high-dimensional vector spaces where similar campaigns cluster near each other based on their dimensional characteristics.

Think of it like this: every campaign gets mapped to a point in a space with dozens of dimensions—one for audience awareness, one for competitive intensity, one for message sophistication, and so on. Campaigns that are dimensionally similar end up near each other in this space, even if their surface characteristics (industry, product, channel) are completely different.

Then you use *similarity metrics*—cosine similarity being the most common—to calculate how close two campaigns are in that dimensional space. High similarity scores indicate rhyming dimensional profiles. Low similarity scores indicate fundamentally different patterns.

The sophisticated part isn't the math—it's determining which dimensions to measure, how to extract them from unstructured campaign descriptions, how to weight them for different analytical questions, and how to validate that calculated similarity actually predicts comparable performance patterns.

That requires several things:

Multi-LLM orchestration

Different models excel at different analytical tasks. Use Claude for strategic reasoning, GPT for creative variation generation, Gemini for data analysis—and ensemble their outputs for better results than any single model can produce.

Structured extraction frameworks

Apply careful prompt engineering to extract dimensional characteristics consistently from diverse campaign descriptions.

Human-in-the-loop (HITL) validation

Experienced marketers review dimensional mappings to ensure AI extracted the right characteristics and weighted them appropriately.

Continuous refinement

Each new campaign analyzed improves the dimensional framework, makes the embeddings more accurate, and strengthens the pattern recognition.

This kind of pattern recognition requires a sophisticated infrastructure. It's not something you build casually with off-the-shelf tools. It requires purpose-built architecture designed specifically for multi-dimensional campaign analysis.

We'll go deeper on the technical implementation in Part 3. For now, just understand this: this approach is made real by AI, but it's made *useful* by experienced human expertise guiding the analysis.

Why 'Always-On' Intelligence Beats Static Benchmarks

There's something fundamentally different about how pattern recognition works compared to traditional benchmarking, and it's worth making explicit: dimensional pattern recognition isn't about finding *the* answer and being done. It's about building a competitive intelligence system that continuously surfaces new opportunities.

Benchmarks give you false certainty; they make it seem like as long as you hit them, your work is done. "We beat the industry average—campaign successful, move on to the next one." It's comfortable. It's definitive. It's also stagnant.

Pattern recognition operates differently. It's an always-on hypothesis engine that continuously generates testable insights as you accumulate more dimensional data across more campaigns.

Here's how this plays out in practice:

Month 1

You run a campaign. Dimensional analysis suggests it rhymes with three previous campaigns across audience context and message strategy dimensions. You test their lessons. Performance improves 23%.

Month 3

You run another campaign. Now you have more dimensional data. Pattern recognition reveals that your Month 1 success wasn't just about message strategy—there was a competitive intensity dimension you hadn't weighted heavily enough. New hypothesis to test.

Month 6

Patterns are emerging across multiple campaigns. That competitive intensity insight actually interacts with buying stage awareness in non-obvious ways. AI surfaces this interaction effect. You test it. Another performance lift.

Month 12

Your dimensional pattern database now contains rich multi-campaign intelligence. You're discovering rhymes your competitors can't see because they're analyzing campaigns in isolation or comparing to static benchmarks.

While being "always on" might initially sound intimidating, this isn't exhausting. Once you develop it as a normal practice, it'll compound your competitive advantage.

Every campaign you run adds dimensional intelligence that makes pattern recognition more sophisticated. Every hypothesis you test—whether it succeeds or fails—provides additional data that refines which dimensional combinations actually predict success.

Your competitors are asking "did we beat the benchmark?" and moving on. You're building an intelligence system that gets smarter with every campaign, discovering advantages that only become visible through multi-dimensional analysis at scale.

This doesn't mean blindly following the AI's suggestions: sometimes AI will suggest dimensional rhymes that turn out not to be predictive. The mathematical similarity is real, but the lessons don't transfer as expected when you test them.

This isn't a flaw in the approach—it's exactly why human expertise is essential in the loop.

When AI suggests a dimensional rhyme that doesn't hold up in practice, experienced humans ask: "Why didn't this transfer? What dimension did we miss or weigh incorrectly? What contextual factor invalidated the pattern?" Those questions refine the analytical framework and make future pattern recognition more accurate.

This is fundamentally different from benchmark optimization, where false positives are invisible. You can beat a benchmark with a campaign that's actually underperforming relative to its true potential, and you'd never know because you're measuring against a statistical fiction.

With dimensional pattern recognition, you're constantly validating hypotheses against reality. Every test provides feedback that improves the system. Failed hypotheses aren't wasted effort—they're learning opportunities that sharpen future pattern recognition.

Think of it like a stock market analyst. The market is constantly changing. New information emerges daily. Yesterday's analysis informs but doesn't determine today's decisions. The analyst with the most sophisticated analytical framework and deepest experience reading patterns will consistently spot opportunities others miss—not because they're always right, but because they're asking better questions and refining their approach continuously.

Pattern recognition in marketing works the same way. The competitive landscape shifts. Buyer behaviors evolve. Channel effectiveness changes. What worked last quarter might not work next quarter—not because the approach was wrong, but because the conditions changed.

Static benchmarks can't adapt to that reality. They're rearview mirrors showing you what happened in the past to campaigns that weren't yours.

Dimensional pattern recognition is forward-looking radar that continuously updates as conditions change, surfaces new hypotheses to test, and builds compounding intelligence that creates sustainable competitive advantage.

You're not searching for certainty. You're building a system that generates strategic advantages faster than your competitors can copy them.

That's not a limitation. That's the entire point.

In the next chapter, we'll explore specific examples of how campaigns rhyme across dimensions—concrete scenarios showing what this looks like in practice and why the lessons transfer in ways traditional benchmarks never reveal.

Then in Chapter 6, we'll build out the complete five-dimension framework, showing exactly how to map campaigns, calculate similarity, and identify the rhyming patterns that matter most for your specific questions.

But for now, sit with this shift in perspective:

Stop asking "what's the benchmark for campaigns like mine?"

Start asking "what campaigns share dimensional signatures with mine?"

The campaigns that rhyme aren't necessarily in your industry, targeting your personas, using your channels, or selling your type of product.

But they succeeded—or failed—for reasons that will inform your strategy in ways no industry average ever could.

That's the fundamental insight. That's the alternative to benchmarks.

And that's what the rest of this book will teach you to recognize, analyze, and leverage.

Chapter Five

How Campaigns Rhyme

In Chapter 4, I introduced the core idea: dimensions rhyme even when campaigns don't.

That's an abstraction. It's a framework. It's intellectually satisfying if you're already convinced.

But it doesn't help you *see* it yet. It doesn't make you recognize the dimensional patterns hiding in your own experience.

So let's get concrete.

This chapter is going to walk you through multiple scenarios describing campaigns that look nothing alike on the surface but share profound dimensional similarities underneath. I'm going to show you what I see when I analyze campaigns, drawing from having led thousands of campaigns over years and noticing which patterns repeat and which lessons transfer across wildly different contexts.

There are four detailed scenarios ahead, each one pairing campaigns that traditional benchmarks would never connect. Each one reveals a different type of dimensional rhyme, and by the end, you'll start recognizing these patterns in your own work.

These aren't case studies with specific client results or performance metrics. They're realistic explorations showing how experienced marketers recognize patterns that industry categories and traditional digital agencies miss entirely.

After leading thousands of campaigns and watching which strategies succeed and fail across dozens of industries, channels, and buyer types, I've developed pattern recognition that operates almost intuitively now. I can see the scene, the emotion, and all the individual dots simultaneously. It's as if I can view a Seurat painting from both far enough away to recognize the image and closely enough to see each distinct point of color that creates it. I'll convey to you how to do this, too. The best way to begin to explain it is through this series of example scenarios. By the end of this chapter, you'll start seeing these dimensional rhymes too.

Let's begin.

Scenario 1: Innovation Seekers in Emerging Categories

Consider two campaigns that would never appear in the same benchmark comparison: a wellness platform and an environmental sustainability software vendor. In both of these cases, however, the target consumer is motivated by a desire to be innovative and a need for material to convince stakeholders to join them in taking a risk.

- **Campaign A:** A productivity startup targeting progressive HR leaders at tech companies with an AI-powered employee wellness platform delivered through promoted LinkedIn thought leadership posts and virtual executive roundtables.
- **Campaign B:** A sustainability software vendor targeting forward-thinking operations VPs at consumer brands with carbon tracking tools promoted through industry conferences and peer case studies.

Completely different categories. Different buyer personas. Different products. Different channels. Different value propositions.

Traditional benchmarks would tell the productivity startup to compare against "HR tech LinkedIn campaigns" and the sustainability vendor to compare against "environmental software conference marketing."

But here's what those benchmarks miss: these campaigns share profound dimensional similarity across the variables that actually predict performance. These variables—audience behavior, competitive environment, message strategy, execution approach, and business goals—are key places to look for rhymes:

Dimension 1: Audience behavior

Both campaigns target executives who are fundamentally innovation-seeking. Neither buyer is motivated by fear or risk mitigation. Both are trying to establish a competitive advantage and internal credibility through early adoption. Both have been frustrated by incremental solutions that don't deliver transformative change. Both face the challenge of selling new ideas internally to skeptical stakeholders. Both need ammunition to champion innovation to conservative leadership teams.

The HR leader isn't buying "employee wellness." They're buying "I'm positioning our company as a progressive employer that attracts top talent." The operations VPs aren't buying "carbon tracking." They're buying "I'm driving sustainability leadership that differentiates our brand."

Same fundamental buyer psychology. Same change-agent positioning. Same internal selling challenge.

Dimension 2: Competitive environment

Both campaigns operate in emerging categories where the competition isn't other vendors—it's the status quo. Both buyers have cobbled together partial solutions using spreadsheets and manual processes. Both face skepticism about whether new technology solves real problems or just adds complexity. Both work in organizations where "we've always done it this way" is the real competitor. Both need to create urgency for problems that aren't yet seen as urgent.

The competitive challenge isn't "beat other wellness platforms" or "beat other carbon trackers." It's "convince forward-thinking executives that now is the time to move, that first-mover advantage matters, that waiting means falling behind progressive competitors."

Dimension 3: Message strategy

Both campaigns succeed or fail based on one critical capability: painting a vision of what's possible without sounding naive about what's difficult.

You can't lead with "transform your workplace overnight" because the HR leader knows change is hard and has seen overpromised initiatives fail. You can't lead with "achieve carbon neutrality easily" because the operations VP understands the complexity and won't trust simplistic claims.

But you also can't acknowledge so much difficulty that you kill enthusiasm for change. These are optimistic buyers who need their ambition validated, not their concerns amplified.

The message strategy that works: vision-focused positioning ("here's where progressive companies are headed"), competitive advantage framing ("early adopters are already seeing these benefits"), and evidence-based optimism ("here's realistic evidence this works, with honest acknowledgment of what's required").

Not fear-driven. Not risk-focused. Ambition-driven change narratives.

Dimension 4: Execution approach

Both campaigns require thought leadership before product consideration. You can't go straight to feature comparisons. You need to establish category vision through educational content that demonstrates where the industry is heading—not generic thought leadership, but specific strategic frameworks that help champions build internal cases for change.

Both campaigns benefit from peer validation through communities and roundtables more than analyst reports. Both need longer education cycles because buyers are creating evaluation criteria, not following established ones. Both perform better with executive-level engagement rather than practitioner-level tactics.

Dimension 5: Business goals

Both campaigns optimize for market education over the immediate pipeline. Both have longer sales cycles (6–12 months) because they're category creation, not replacement purchases. Both involve internal evangelism and stakeholder alignment. Both require executive championship. Both measure success by market positioning, not just revenue.

Traditional benchmarks do provide some information. If you're running the productivity startup campaign and you compare yourself to "HR tech LinkedIn benchmarks," you'll learn generic optimization tactics—best times to post, optimal content length, image vs. video performance.

But if you recognize the dimensional similarity to the sustainability campaign, you learn strategic insights:

- Lead with vision and competitive advantage, not features or fear.

- Emphasize first-mover benefits in your messaging.

- Build communities of progressive practitioners, not just prospect lists.

- Invest in category creation content over product marketing.

- Optimize for executive championship metrics, not just lead volume.

These strategic insights transfer because the dimensional signatures rhyme—even though "HR tech" and "sustainability software" share no benchmark category.

The campaigns look completely different on the surface. The dimensions underneath rhyme perfectly.

And those dimensional rhymes predict which strategic approaches will work.

Scenario 2: Challenger Positioning Across Contexts

Let's examine two more campaigns that traditional benchmarks would never connect: a new project management platform and a familiar company introducing a new data analytics feature. Both campaigns need to convince their audience to try a new product, even though there are already solutions available.

Campaign A: A late-stage startup with a project management platform running LinkedIn ads targeting IT directors at mid-market companies, positioned against entrenched incumbent leaders.

Campaign B: An established vendor launching a new data analytics module aimed at existing customers who currently use competitive point solutions, using email and webinar campaigns.

Different channels. Different audience relationships. Different competitive contexts. Different business stages.

But watch what happens when you map them dimensionally.

Dimension 1: Audience behavior

Both campaigns target buyers who already have a solution. In Campaign A, IT directors use established project management tools (Asana, Monday, Jira) and face switching costs. In Campaign B, existing customers use third-party analytics tools and face integration complexity.

Neither buyer woke up looking for a new solution. Both need to be convinced that the friction of change is worth the benefit of switching. Both have existing workflows built around current tools. Both face internal resistance ("Why are we changing something that works?").

The buyer psychology isn't "find the best tool"; it's "is this worth the disruption?"

Dimension 2: Competitive environment

Both face incumbent inertia, not just incumbent products. Campaign A fights Asana's market leadership and ecosystem. Campaign B fights internal integration investments in existing analytics tools.

But here's the dimensional rhyme: both campaigns succeed by positioning themselves as "the challenger who understands what the incumbent gets wrong." Not "we're better at what they do," but "they're optimizing for the wrong things, and here's why that matters."

The late-stage startup can't outspend Asana on features or marketing. Instead of relying on flashy offerings, they need to reframe what "better" means. The established vendor can't position their analytics module as "more powerful" than dedicated analytics platforms. They need to reframe the decision around integration efficiency and unified workflows.

Dimension 3: Message strategy

Both campaigns use challenger positioning frameworks. Both emphasize what the incumbent approach misses rather than feature-by-feature comparison. Both use customer evidence of "wish I'd switched sooner" rather than "this has more features."

For Campaign A, the message isn't "we have better project management features." It's "your current tool optimizes for task tracking, but what you actually need is workflow intelligence and a different architectural approach."

For Campaign B, the message isn't "we have better analytics." It's "you're cobbling together point solutions that break every time you update something, and unified architecture solves that problem permanently."

Different specifics. Same strategic framing. Same message architecture.

Dimension 4: Execution approach

Both campaigns benefit from competitive comparison content (vs. incumbent solutions) more than generic product marketing. Both perform better with longer-form content that substantiates the strategic reframing. Both need to demonstrate customer success stories focused on "why we switched" rather than "what we achieved."

Campaign A's LinkedIn ads drive to comparison landing pages, not generic product pages. Campaign B's webinars focus on "rethinking analytics strategy" rather than "here's our new module."

Dimension 5: Business goals

Both campaigns optimize for conversion quality over volume. Both have moderate sales cycles (3–6 months) involving evaluation committees. Both measure success through competitive displacement metrics—tracking not just closed revenue but specifically how many deals came from competitor replacements.

Now here's what those dimensional rhymes reveal:

If Campaign A compares itself to "LinkedIn ad benchmarks for project management software," they'll optimize click-through rates and cost-per-lead without ever learning the strategic insight that matters most: challenger positioning content outperforms feature-focused content by 3x in comparable situations.

If Campaign B compares itself to "email benchmarks for existing customer upsell," they'll optimize open rates without discovering that long-form competitive comparison webinars convert 5x better than product demos when you're displacing incumbent solutions.

But recognize the dimensional rhyme—both are challenger positioning plays targeting buyers with incumbent inertia—and suddenly Campaign A's LinkedIn strategy informs Campaign B's webinar content strategy. Campaign B's customer success story framing informs Campaign A's landing page messaging.

The insight isn't "LinkedIn ads work" or "webinars work." It's "challenger positioning content works when you're fighting incumbent inertia, regardless of channel."

That's a dimensional rhyme. And it's the kind of insight that traditional benchmarks, which compare LinkedIn ads to LinkedIn ads and webinars to webinars, will never reveal.

Scenario 3: The Hidden Rhyme of Buyer Sophistication

Here's a dimensional pattern that most marketers would miss entirely because the surface characteristics look so different: for both a workforce scheduling software platform and a financial planning platform, the buyer is a technically sophisticated consumer looking for detailed specifics about the product.

- **Campaign A**: An HR software vendor targeting IT directors at mid-market retailers with email campaigns about workforce scheduling optimization.
- **Campaign B**: A financial planning platform targeting CFOs at mid-market professional services firms with display advertising about cash flow forecasting.

Ask any benchmark which industry category to compare against, and you'll get: "HR software email campaigns" and "financial software display campaigns."

But look at the dimensional signature:

Dimension 1: Audience behavior

Both target technically sophisticated buyers with specific domain expertise. The IT director understands software architecture, integration complexity, and implementation risks. The CFO understands financial modeling, forecasting methodologies, and reporting requirements.

Neither is a naive buyer who needs education on their domain. Both are skeptical of vendor claims because they've seen inflated promises before. Both will personally evaluate technical capabilities before involving procurement. Both have strong opinions about what "good" looks like in their domain.

More importantly: both are evaluating you against their own internal expertise, not against what other vendors say. The IT director knows workforce scheduling algorithms. The CFO knows financial modeling. They're not comparing you to competitors—they're comparing you to their own understanding of what's possible.

Dimension 2: Competitive environment

Both operate in markets where buyer sophistication creates specific competitive dynamics. Generic positioning doesn't work. "Best-in-class workforce scheduling" means nothing to an IT director who can assess algorithmic sophistication themselves. "Comprehensive cash flow forecasting" means nothing to a CFO who understands statistical modeling.

Both buyers filter out generic marketing immediately. Both look for technical depth markers that signal you actually understand their domain. Both respond to specificity over claims.

Dimension 3: Message strategy

Both campaigns must demonstrate domain credibility before product consideration. You can't lead with benefits ("optimize your workforce scheduling!") because sophisticated buyers tune out benefit claims instantly. You need to lead with evidence of technical depth.

For the HR software: mention specific scheduling constraint variables, reference algorithmic approaches, demonstrate understanding of retail-specific workforce dynamics (not generic "workforce management").

For the financial platform: reference specific forecasting methodologies, show understanding of professional services revenue recognition complexity, demonstrate grasp of cash conversion cycle nuances.

The message strategy isn't "here's why this matters" (they already know why it matters). It's "here's evidence that we understand this at your level of sophistication."

Dimension 4: Execution approach

Both campaigns perform better with technical content (whitepapers, technical documentation, detailed methodology explanations) than with typical marketing content (customer testimonials, benefit claims, generic case studies).

Both need longer-form content that demonstrates depth. Both benefit from free trials approaches where the buyer can evaluate capabilities themselves. Both perform worse with aggressive follow-up and better with consultative approaches.

Dimension 5: Business goals

Both campaigns optimize for deal size over volume. Both have longer sales cycles (4–9 months) dominated by technical evaluation phases. Both involve technical buyers who influence but don't always approve budgets. Both measure success by average contract value, not lead volume.

Thinking in terms of these five dimensions provides a crucial insight.

These campaigns share nothing in terms of industry category, channel choice, or product type. "HR software email benchmarks" and "financial software display benchmarks" would never suggest comparing them.

But they share profound similarity in buyer sophistication levels—and that dimensional similarity predicts strategic success patterns:

- Technical depth content outperforms benefit-focused content.

- Longer-form educational content converts better than short-form promotional content.

- Consultative follow-up approaches work better than aggressive sales development.

- A free trial offer outperforms demo requests or gated content.

- Technical documentation downloads predict deal size better than webinar attendance.

If Campaign A compares itself to "HR software benchmarks," they might conclude that email subject line optimization and promotional offers are the path to improvement.

But recognize the dimensional rhyme with Campaign B—buyer sophistication creates specific strategic requirements—and they'd learn that investing in technical whitepapers, detailed methodology documentation, and free trial programs yields better results than subject line testing.

The campaigns rhyme across a dimension (buyer sophistication) that doesn't exist in any benchmark category. But that dimensional rhyme predicts performance patterns more reliably than industry categories ever could.

Scenario 4: When Dimensions Contradict Categories

Here's where dimensional pattern recognition gets really powerful—when it reveals insights that directly contradict what industry benchmarks suggest. We'll look at two campaigns that seem very similar, but that have important differences under the surface. Two HR software vendors are targeting different consumers: one with big-picture concerns, and the other with a more fine-grained, technical focus.

Campaign A: An HR software vendor targeting chief human resources officers at enterprise healthcare systems with thought leadership content and executive roundtable events.

Campaign B: An HR software vendor targeting IT directors at mid-market retail companies with email campaigns about API capabilities and technical implementation guides.

Same product category. Same "HR software" industry classification. Traditional benchmarks would compare both against "HR software campaigns" and suggest they should use similar strategies.

But map them dimensionally, and you discover they share almost nothing:

Campaign A targets senior executives (CHROs) focused on strategic workforce initiatives. They evaluate based on organizational change management, executive peer validation, analyst endorsements, and strategic vision. They have large budgets and long buying cycles (12+ months), and care about vendor thought leadership positioning. The decision is strategic, not technical.

Campaign B targets technical evaluators (IT directors) focused on implementation complexity. They evaluate based on API documentation quality, integration specifications, technical support capabilities, and implementation timelines. They have constrained budgets, moderate buying cycles (4–6 months), and care about technical credibility. The decision is technical, not strategic.

These campaigns share a product category ("HR software") but differ across nearly every other dimension: buyer role, buying criteria, decision-making authority, evaluation process, content preferences, and success metrics.

"HR software benchmarks" would suggest they should use similar strategies. But dimensional analysis reveals they need opposite approaches:

Campaign A needs the following:

- Thought leadership content (industry trend analysis, strategic vision pieces)

- Executive events (roundtables, advisory boards, conference speaking)

- Long-form strategic content (research reports, industry studies)

- CEO/founder visibility and executive relationships

- Brand-level positioning and analyst relations

Campaign B needs the following:

- Technical documentation (API guides, integration specifications)

- Product education content (implementation guides, technical webinars)

- Hands-on evaluation programs (free trials, sandbox environments)

- Technical support responsiveness and developer relations

- Product-level capabilities and technical credibility

The dimensional signatures don't rhyme at all—despite sharing an industry category.

In fact, Campaign B has more in common with a developer tools campaign targeting IT directors (completely different product category) than it does with Campaign A (same product category).

This is where dimensional thinking breaks you free from category-based benchmarks entirely.

Industry categories are convenient labels. They're not predictive of strategic success patterns. Dimensional signatures are predictive.

What You're Learning to See

By now, you should be noticing something: campaigns that look nothing alike on the surface can share deep dimensional patterns. And campaigns that look similar on the surface can differ across every dimension that matters.

The cybersecurity campaign and the compliance campaign rhymed across buyer psychology dimensions. The project management startup and the analytics module launch rhymed across competitive positioning dimensions. The workforce scheduling campaign and the financial forecasting campaign rhymed across buyer sophistication dimensions. And the two HR software campaigns—supposedly "comparable" by industry standards—rhymed across zero meaningful dimensions.

This is what dimensional pattern recognition looks like in practice.

It's not about finding campaigns in your industry. It's not about finding campaigns in your channel. It's not about finding campaigns at your company stage or targeting your personas.

It's about finding campaigns that share dimensional signatures across the variables that actually predict performance patterns.

And here's what makes this approach so powerful: the insights transfer because the dimensional similarities create comparable strategic contexts—even when every surface characteristic differs.

When you recognize that a campaign rhymes with yours across buyer psychology dimensions, you can transfer message strategy insights. When dimensions rhyme across competitive environment variables, you can transfer positioning approaches. When dimensions rhyme across execution complexity, you can transfer channel strategy and nurture sequence designs.

The rhyming dimensions tell you what lessons are transferable and why they transfer.

Traditional benchmarks can't do this. They can only tell you "here's what everyone else in your category did" without any framework for understanding which aspects of those campaigns are relevant to yours.

Dimensional pattern recognition gives you the framework for understanding *why* certain campaigns are comparable and *which* strategic elements transfer.

The Uncomfortable Question

But this creates a challenge:

How do you actually *find* these dimensional rhymes at scale?

You've just read four scenarios where I showed you the patterns. But I've been doing this for 25+ years, having led thousands of campaigns across dozens of industries. I've seen these dimensional patterns repeat so many times across such diverse contexts that I recognize them intuitively now—the way an experienced doctor can diagnose conditions based on subtle symptom combinations that medical students wouldn't notice.

You can't manually map campaigns across dozens of dimensions and calculate similarity scores in a spreadsheet. Your brain can't process that level of multidimensional complexity at scale. And even if you could, you'd be limited to your own direct experience—which might represent dozens or even hundreds of campaigns, but without the dimensional diversity needed to recognize which patterns actually predict transferable insights versus which are just coincidental correlations.

Traditional digital agencies can't do this either. They have the raw material—multiple clients, diverse campaigns, dimensional variety—but they organize their teams by client account, keeping insights locked in separate silos. The team working on healthcare campaigns never connects their learnings to the team working on financial services campaigns, even when those campaigns share rhyming dimensions around buyer psychology or competitive positioning. They lack both the organizational structure and the analytical framework to extract dimensional patterns across their client portfolio.

This is where most marketers get stuck. They intellectually understand that dimensional pattern recognition is better than industry benchmarks. They can see the dimensional rhymes when someone points them out. But they don't have a practical method for systematically finding those patterns, and they don't have access to the diversity of campaign data needed to separate signal from noise.

And that's exactly the gap that sophisticated pattern recognition capabilities fill—combining dimensional analysis frameworks with experience across thousands of campaigns to identify which dimensional signatures actually predict transferable insights.

The next chapter will explore the five core dimensions that matter most for campaign pattern recognition, showing you what dimensions create meaningful rhymes and why certain dimensional combinations predict strategic success more reliably than others. But I won't be providing a step-by-step implementation guide. Instead, I'll provide you with a fundamental understanding of how to find your own rhymes, so you can apply the method to your own diverse set of campaigns and generate reliable insights.

Then, in Chapter 7, we'll explore why finding these dimensional patterns requires both AI capabilities (to process the complexity at scale) and deep human expertise (to know which patterns actually matter and how to interpret what they mean strategically).

But before we get there, I want you to sit with what you've learned in this chapter.

Think about campaigns you've led in the past. Think about what worked and what didn't. Now try thinking about them dimensionally:

- What was the audience behavior at play? (Risk-averse? Ambitious? Skeptical? Sophisticated?)

- What was the dynamic of the competitive environment? (Incumbent displacement? Category creation? Crowded market?)

- What was your message strategy? (Thought leadership? Technical depth? Fear reduction? Aspirational gains?)

- What was the execution approach? (Long nurture? Short burst? Account-based? Broad demand gen?)

- What were the business goals? (Pipeline quality? Revenue velocity? Market share? Brand awareness?)

Now try to think about other campaigns—in different industries, different channels, different contexts—that might share dimensional signatures with your most successful initiatives.

Can you identify them? Can you articulate why they rhyme? Can you extract the strategic lessons that would transfer?

If you can, you're starting to think dimensionally.

If you can't—or if you can see the pattern but don't have access to enough diverse campaign data to validate whether it's a real signal or just a coincidence—you're discovering exactly why dimensional pattern recognition requires both sophisticated analytical frameworks and extensive campaign experience to deliver reliable insights.

What makes this possible today—and what wasn't possible even five years ago—is the combination of sophisticated AI capabilities with experienced marketing judgment. The AI provides computational power to analyze patterns across thousands of campaigns and dozens of dimensions simultaneously. The human expertise provides strategic judgment to know which patterns actually matter, how to weight them contextually, and what they mean strategically. Neither works alone at scale.

Those are the campaigns you should be learning from.

Not the ones in your industry benchmark report.

The ones that rhyme.

Chapter Six

The Five Dimensions

IN CHAPTER 5, YOU saw dimensional pattern recognition in action—four scenarios where campaigns that looked nothing alike on the surface shared deep structural similarities underneath.

A cybersecurity campaign rhymes with a compliance campaign because both target risk-averse executives making fear-driven decisions. A project management startup rhymes with an established vendor's product launch because both fight incumbent inertia. An HR software campaign rhymes more with a finance software campaign than with other HR software campaigns because buyer sophistication matters more than product category.

Those weren't cherry-picked examples. They're representative of what happens when you stop thinking categorically and start thinking dimensionally.

But to see those patterns consistently—to recognize them in your own campaigns before running them, not just in retrospect—you need to understand what dimensions actually matter and why certain dimensional combinations create predictive patterns while others just add noise.

That's what this chapter explores.

I'm going to walk you through the five core dimension categories that, across thousands of campaigns spanning 25+ years, I've found consistently predict whether strategic insights will transfer from one campaign to another. These aren't arbitrary classifications I invented. They're patterns that emerge when you analyze enough campaigns to separate signal from noise—the dimensions that actually correlate with performance outcomes rather than just convenient ways to organize spreadsheets.

I am also providing some examples of components within each of these five core dimensions. These sub-dimensions are by no means a complete list, but they are illustrative of the types of qualitative classifications that make pattern recognition actionable by smart marketers.

This chapter explains the dimensional framework. Chapter 7 will explore why this human-AI partnership creates a defensible competitive advantage that actually appreciates as AI advances. But first, you need to understand what dimensions matter and why.

Let's begin.

Dimension 1: Audience Behavior

The first dimension—and often the most predictive—is understanding who you're actually talking to and what psychological state they're in when they encounter your campaign.

This isn't demographic data. Company size, industry, job title—those are labels, not psychological insights. A CISO at a 500-person healthcare company and a CISO at a 500-person fintech company might share a job title and company size, but they can exist in completely different psychological universes when they are evaluating security solutions.

What matters is the deeper pattern of how they think, decide, and behave.

Buyer Sophistication

Some buyers are encountering your problem space for the first time. Others have tried and failed with three previous solutions. Some have deep domain expertise and can evaluate technical details. Others are generalists who need to rely on peer validation and external proof.

These aren't better or worse buyers—they inhabit different psychological contexts requiring fundamentally different campaign approaches.

Consider technical sophistication variance in marketing technology (martech) campaigns. A CMO who started as a software engineer evaluates martech platforms by reading API documentation, testing webhook reliability, and assessing data architecture.

A CMO whose career progressed through brand marketing evaluates the same platforms by examining campaign examples, reviewing agency partnerships, and assessing ease of use for non-technical teams.

Same title. Same budget authority. Same strategic objective. Completely different evaluation criteria based on technical sophistication.

The technically sophisticated CMO's campaign rhymes with developer tools marketing—documentation quality matters more than testimonials, architectural diagrams resonate more than ROI calculators, and integration capabilities outweigh interface design. The campaign needs to prove technical competence before business value.

The brand-background CMO's campaign rhymes with professional services marketing—case study quality matters more than technical specifications, peer testimonials resonate more than feature lists, partnership ecosystems outweigh API flexibility. The campaign needs to prove business outcomes before technical capabilities.

Traditional benchmarks can't capture this. "Martech email campaigns to CMOs" averages the engineer-CMO and the brand-CMO into meaningless noise. Dimensional pattern recognition separates them into distinct psychological profiles that predict which strategic approaches will work.

Decision Authority

Who makes the decision, who influences it, and who can kill it are three different psychological dynamics that fundamentally change campaign strategy.

An economic buyer with budget authority and profit and loss (P&L) responsibility thinks differently than a technical evaluator who can recommend but not approve. An executive sponsor who champions your solution internally faces different political dynamics than an end-user advocate who loves your product but lacks influence with leadership.

These dynamics create dimensional patterns that matter more than job titles.

A campaign targeting economic buyers with clear authority rhymes with other economic-buyer campaigns—regardless of industry or product—because the psychological pattern of ROI focus, risk calculation, and stakeholder management transfers. The messaging emphasizes business outcomes over features, proof points over innovation, and de-risking over disruption.

A campaign targeting technical evaluators without budget authority rhymes with other technical-evaluator campaigns because the psychological pattern of thoroughness, skepticism toward marketing claims, and peer validation transfers. The messaging emphasizes technical depth over business benefits, architectural fit over ROI, and expert credibility over executive testimonials.

A campaign targeting consensus committees where no single person has authority rhymes with other consensus-committee campaigns because the psychological pattern of risk distribution, lowest-common-denominator decision-making, and political navigation transfers. The messaging emphasizes safe choices over bold moves, category leadership over innovation, and broad stakeholder satisfaction over departmental optimization.

Same company size, same industry, same channel—but three completely different dimensional profiles requiring three completely different strategic approaches.

Risk Tolerance

Some buyers actively seek innovation and competitive advantage through early adoption. Others avoid risk and prefer validated, safe choices. Some are in organizational contexts that reward bold moves. Others face severe consequences for failed initiatives.

This psychological dimension predicts campaign strategy more reliably than almost any demographic variable.

Risk-averse buyers in consequence-heavy environments need campaigns that emphasize safety, validation, and de-risking. They respond to peer proof ("companies like yours"), category leadership signals ("trusted by 10,000+ organizations"), and risk mitigation messaging ("seamless implementation" and "enterprise support"). They're skeptical of innovation claims and disruption narratives.

Risk-tolerant buyers in innovation-rewarding environments need campaigns that emphasize competitive advantage, forward-thinking positioning, and strategic differentiation. They respond to vision statements, early adopter communities, and cutting-edge capabilities. They're skeptical of "safe choice" positioning and "nobody gets fired for buying us" messaging.

These are fundamentally different psychological profiles that predict campaign performance independently of every other variable.

A cybersecurity campaign targeting risk-averse CISOs at regulated financial institutions rhymes with a compliance campaign targeting risk-averse general counsels at healthcare providers—not because they're both "compliance-related," but because the psychological pattern of fear-driven decision-making, consequence-heavy environments, and safe-choice optimization transfers perfectly.

A developer tools campaign targeting risk-tolerant engineering leaders at fast-growing startups rhymes with a business intelligence campaign targeting risk-tolerant CMOs at digital-native brands—not because they're both "technology buyers," but because the psychological pattern of innovation-seeking, competitive-advantage focus, and early-adopter enthusiasm transfers.

Awareness Stage

Where buyers are in their journey from being unaware of the problem to evaluating the solution fundamentally changes what campaigns can accomplish and how messages must be structured.

Problem-unaware buyers don't yet recognize they have an issue worth solving. They need education, pattern recognition, and "aha moments" that surface latent pain. Campaigns targeting them focus on symptoms ("are you experiencing X?"), consequences ("here's what X costs you"), and reframing ("what you think is normal is actually fixable").

Problem-aware but solution-unaware buyers recognize the issue but don't know solutions exist. They need category education, proof of concept, and validation that solving this problem is feasible. Campaigns targeting them focus on possibility ("you can actually solve X"), proof points ("here's how others solved it"), and approach validation ("here's why this method works").

Buyers who are aware of the solution but are still evaluating vendors understand the category and are comparing options. They need differentiation, competitive positioning, and proof of superiority. Campaigns targeting them focus on why you're better ("here's what makes us different"), proof of results ("here's what we delivered"), and trust signals ("here's why you should believe us").

These are completely different campaign objectives requiring completely different strategic approaches—yet traditional benchmarks average them all together into "industry email performance."

A campaign creating problem awareness in a new category rhymes with other awareness-creation campaigns, regardless of industry, because the psychological pattern of surfacing latent pain, overcoming status quo bias, and creating urgency transfers. The execution might involve thought leadership content, research reports, or diagnostic tools—but the strategic pattern is consistent.

A campaign differentiating in a crowded, established category rhymes with other competitive-differentiation campaigns regardless of product type, because the psychological pattern of cutting through noise, establishing credible superiority claims, and overcoming incumbent inertia transfers. The execution might involve comparison content, customer success stories, or analyst validation, but the strategic pattern is consistent.

The Interaction Effects

Here's where dimensional thinking gets sophisticated: these audience psychology variables don't operate independently. They interact and combine to create distinct psychological profiles.

A risk-averse, experienced buyer with decision authority evaluating solutions in a mature category creates one dimensional profile. A risk-tolerant, first-time buyer with influence, but not authority, exploring solutions in an emerging category creates a completely different profile.

Those two buyers could have identical job titles at identically sized companies in the same industry, but campaigns that work for one will fail for the other because their psychological profiles are completely different.

Traditional benchmarks can't capture this complexity. They reduce everything to simple categories and average out the nuance. Dimensional pattern recognition preserves the complexity and uses it to find genuinely predictive patterns.

When you map campaigns across these audience psychology dimensions, you start seeing which psychological profiles cluster together and which campaigns addressing similar profiles delivered similar results regardless of surface differences.

That's not guessing. That's pattern recognition.

And audience psychology is just the first dimension.

Dimension 2: Competitive Environment

The second dimension captures the strategic environment your campaign operates within—the competitive dynamics, market maturity, and positioning challenges that shape what's possible and what approaches will work.

Two campaigns can target psychologically identical buyers but require completely different strategies because of where they sit in competitive and market contexts.

Market Maturity

Emerging categories with low awareness require fundamentally different campaign strategies than mature categories with established vendors and buyer expectations.

In emerging categories, buyers don't have mental models for evaluating solutions. They don't know what features matter, what questions to ask, or what constitutes "good performance." They're creating evaluation frameworks from scratch, often by analogy to other categories they understand better.

Campaigns in emerging markets focus on category education, creating evaluation criteria, and establishing reference points. Success comes from shaping how buyers think about the problem space—defining what matters before competitors can establish different criteria. The message isn't "we're better than alternatives" because buyers don't yet have alternatives in mind. It's "here's how to think about solving this problem."

In mature categories, buyers have established expectations, comparison frameworks, and often negative experiences with previous vendors. They're sophisticated evaluators who've seen the pitch before and developed skepticism toward claims.

Campaigns in mature markets focus on differentiation within established frameworks, overcoming skepticism with proof, and displacing incumbents. Success comes from working within existing mental models while demonstrating superiority on dimensions buyers already care about. The message is explicitly competitive: "here's why we're better than what you're using now."

A campaign launching a new category of workplace communication tools rhymes with other category-creation campaigns—whether in sales intelligence, data visualization, or workflow automation—because the strategic pattern of creating evaluation frameworks, educating skeptical buyers, and establishing reference points transfers across contexts.

A campaign challenging established players in a crowded customer relationship management (CRM) market rhymes with other incumbent-displacement campaigns—whether in marketing automation, project management, or customer support—because the strategic pattern of breaking through noise, overcoming switching costs, and demonstrating superior value transfers.

Competitive Intensity

How many competitors exist, how similar they are, and how noisy the market is fundamentally shapes campaign strategy.

In markets with few clear competitors, campaigns can focus on category benefits and problem-solution messaging. The challenge isn't distinguishing yourself from alternatives—it's convincing buyers the category matters and your approach works.

In crowded markets with dozens of similar-seeming vendors, campaigns must lead with differentiation. The challenge isn't proving the category works; buyers already believe that. It's proving you're meaningfully different from the ten other vendors they're evaluating.

This competitive intensity dimension predicts campaign strategy independently of everything else.

A sales intelligence platform entering a market with 40+ similar vendors faces the same strategic challenge as a marketing automation platform entering a market with 50+ similar vendors, even though they're different product categories serving different personas. Both need campaigns that lead with sharp differentiation, stake out defensible positioning, and give buyers clear reasons to choose them over cheaper/easier/safer alternatives.

The campaigns rhyme because the competitive context creates similar strategic imperatives.

A workflow automation platform creating a new category with minimal direct competition faces similar strategic challenges to those of a business intelligence platform creating a new category, even though they solve different problems for different buyers. Both need campaigns that legitimize the category, create evaluation frameworks, and establish themselves as category leaders before competitors emerge.

The campaigns rhyme because the competitive vacuum creates similar strategic opportunities.

Positioning

Your position relative to market leaders and buyer expectations is relevant to everything about your campaign strategy and message framing.

Challenger positioning—when you're fighting against established incumbents and buyer inertia—requires campaigns focused on "why change," "what you're missing," and "why now." The psychological pattern is disruption, dissatisfaction, and competitive advantage through switching.

Incumbent positioning—when you're defending market leadership against upstarts—requires campaigns focused on "why stay," "what you'd lose," and "why risk it." The psychological pattern is validation, risk reduction, and smart conservatism.

Alternative positioning—when you're establishing a different approach rather than directly competing—requires campaigns focused on "why rethink," "what's possible," and "why this way." The psychological pattern is reframing, possibility, and strategic differentiation.

These positioning contexts predict campaign strategy more reliably than product categories do.

A project management startup challenging Microsoft Project rhymes with a CRM startup challenging Salesforce—not because they're both "software," but because the strategic pattern of displacing entrenched incumbents transfers. Both need campaigns that surface incumbent pain points, validate switching concerns, and provide clear migration paths. Both need messaging that acknowledges the incumbent's strengths while demonstrating where they fall short.

An established marketing automation platform defending against dozens of newer alternatives rhymes with an established CRM tool defending against specialized competitors—not because they're related categories, but because the strategic pattern of incumbent defense transfers. Both need campaigns that emphasize stability, proven results, and the risks of switching to unproven alternatives. Both need messaging that reframes competitor "innovation" as instability and "specialization" as limitation.

A developer platform establishing a new architectural approach rhymes with a data platform establishing a new infrastructure pattern—not because they're in the same stack, but because the strategic pattern of alternative positioning transfers. Both need campaigns that reframe existing solutions as obsolete approaches rather than direct competition. Both need messaging that establishes new evaluation criteria where they win by definition.

Budget Environment

Whether buyers are operating in expansion mode or contraction mode, whether they have discretionary budget or require CFO approval, whether they're replacing existing spend or requesting net-new investment—these financial dynamics shape campaign strategy profoundly.

Campaigns targeting expansion-mode buyers with discretionary budgets can focus on opportunity and growth messaging. The psychological pattern is investment, strategic advantage, and getting ahead of competitors.

Campaigns targeting contraction-mode buyers with scrutinized budgets must focus on efficiency, cost reduction, and replacing existing spend. The psychological pattern is justification, risk mitigation, and doing more with less.

These are fundamentally different strategic contexts that require different campaign approaches regardless of product category.

A productivity software campaign targeting buyers replacing existing tools rhymes with an analytics software campaign targeting buyers consolidating vendors—not because they're both "software," but because the strategic pattern of cost-justified switching transfers. Both need campaigns that demonstrate clear ROI, emphasize replacing existing spend rather than requesting new budget, and position the change as financially prudent rather than risky.

A strategic initiative campaign targeting buyers with discretionary innovation budgets rhymes with other strategic investment campaigns regardless of category, because the psychological pattern of growth orientation, competitive advantage seeking, and forward-looking thinking transfers.

The Contextual Complexity

Just like audience psychology dimensions interact, competitive and market context dimensions combine to create distinct strategic environments.

An emerging category with minimal competition and expansion-budget buyers creates one strategic context. A mature category with intense competition and cost-scrutinizing buyers creates a completely different context.

Campaigns that succeed in the first context often fail catastrophically in the second—not because they're poorly executed, but because the strategic environment is fundamentally different.

Traditional benchmarks ignore this complexity. They compare campaigns across wildly different competitive and market contexts and call it "industry performance."

Dimensional pattern recognition maps these context variables and identifies campaigns operating in similar strategic environments—regardless of whether they share industry labels, product categories, or target personas.

That's when you find insights that actually transfer.

Dimension 3: Message Strategy

The third dimension captures how you're actually communicating—the strategic choices about what to emphasize, how to structure persuasion, and what psychological levers to pull.

This isn't about creative execution or specific copy. It's about the fundamental strategic approach to persuasion that underlies the campaign.

Core Promise

Some campaigns promise achieving positive outcomes—growth, efficiency, competitive advantage, innovation, success. Others promise preventing negative outcomes—risk mitigation, compliance, security, stability, avoiding failure.

These represent fundamentally different psychological appeals that predict campaign strategy independently of product category.

Gain-focused campaigns emphasize opportunity, ambition, and forward momentum. The emotional tone is aspirational. The proof points highlight success stories and achievements. The objection handling focuses on "why miss this opportunity" rather than "why risk this change."

Loss-avoidance campaigns emphasize protection, safety, and risk reduction. The emotional tone is cautious. The proof points highlight disaster avoidance and peace of mind. The objection handling focuses on "what happens if you don't" rather than "what you'll gain if you do."

A cybersecurity campaign focused on threat prevention rhymes with a compliance campaign focused on regulatory risk—not because they're both "security-related," but because the strategic pattern of loss-avoidance messaging transfers. Both lead with fear reduction rather than aspiration. Both use "what could go wrong" framing rather than "what you'll achieve." Both emphasize protection over progress.

A growth marketing platform campaign focused on revenue expansion rhymes with a sales intelligence campaign focused on competitive advantage—not because they're related tools, but because the strategic pattern of gain-focused messaging transfers. Both lead with opportunity rather than risk. Both use "what you'll achieve" framing rather than "what you'll avoid." Both emphasize progress over protection.

These message strategies require different content types, different proof strategies, different emotional tones—yet traditional benchmarks lump them together as "B2B SaaS email campaigns."

Positioning Approach

Some campaigns lead with ideas, frameworks, and intellectual positioning. Others lead with results, proof points, and demonstrated success.

This isn't about whether thought leadership exists in the campaign—it's about whether ideas or proof provide the primary persuasive foundation.

Thought leadership positioning works when buyers are sophisticated enough to evaluate frameworks, when differentiation comes from how you think rather than what you've achieved, or when you're creating new categories where proof is inherently limited. The campaign focuses on teaching, reframing problems, and establishing intellectual authority.

Proof-driven positioning works when buyers are skeptical of claims, when established categories make demonstrations of superiority essential, or when you're displacing incumbents who've already earned trust. The campaign focuses on case studies, metrics, and tangible evidence of results.

A management consulting campaign built around proprietary frameworks rhymes with a research firm's campaign built around unique methodologies—not because they're the same service, but because the strategic pattern of intellectual-authority positioning transfers. Both lead with "here's how we think about this problem differently." Both establish credibility through ideas before results. Both attract buyers who value sophisticated thinking.

An enterprise software campaign built around customer success metrics rhymes with a managed services campaign built around service-level agreement (SLA) commitments—not because they're related offerings, but because the strategic pattern of proof-first positioning transfers. Both lead with "here's what we delivered." Both establish credibility through results before philosophy. Both attract buyers who value demonstrated outcomes.

These positioning approaches require different content strategies, different sales enablement, and different proof architectures, yet they get averaged together in industry benchmarks.

Objection Strategy

How campaigns address buyer concerns reveals fundamental strategic choices about persuasion.

Some campaigns directly address objections—"you think this is too expensive, but here's the ROI" or "you think implementation is hard, but here's how we make it easy." This approach acknowledges concerns and systematically counters them with evidence and reassurance.

Other campaigns reframe objections—"you think price matters, but you're actually choosing between expensive problems and expensive solutions" or "you think implementation complexity is the risk, but the real risk is continuing with inadequate systems." This approach shifts the decision framework rather than arguing within it.

Direct objection handling works when concerns are rational, evidence-based, and soluble through information. Reframing works when concerns reflect deeper mindset issues that information alone can't solve.

A premium-priced solution campaign using reframing ("the question isn't whether we're expensive, it's whether you can afford to keep losing money to inefficiency") rhymes with other premium positioning campaigns regardless of category, because the strategic pattern of shifting evaluation frameworks transfers.

A complex implementation campaign using direct handling ("here's our 90-day deployment plan with dedicated support") rhymes with other complexity-acknowledgment campaigns regardless of product—because the strategic pattern of systematic reassurance transfers.

Emotional Weight

Some campaigns lean heavily into emotional appeals—aspiration, fear, frustration, urgency, belonging. Others lean heavily into rational appeals—logic, data, analysis, frameworks, evidence.

Most campaigns contain both, but the balance point predicts the best strategy.

Emotionally weighted campaigns work when decisions are fundamentally psychological—driven by fear, ambition, identity, or tribal dynamics. They resonate with buyers making decisions based on how options make them feel rather than just what metrics show.

Rationally weighted campaigns work when decisions are fundamentally analytical—driven by clear success criteria, quantifiable outcomes, and logical evaluation frameworks. They resonate with buyers who need to justify decisions with data.

A workplace culture campaign emphasizing belonging and purpose rhymes with a brand positioning campaign emphasizing identity and values—not because they're related categories, but because the strategic pattern of emotion-driven persuasion transfers.

An analytics platform campaign emphasizing data-driven decision-making rhymes with a business intelligence campaign emphasizing quantitative insights—not because they're competitive alternatives, but because the strategic pattern of logic-driven persuasion transfers.

The Strategic Combinations

Message strategy dimensions interact to create distinct persuasive approaches.

A gain-focused, thought-leadership, reframing campaign with emotional weight creates one strategic pattern. A loss-avoidance, proof-driven, direct-objection campaign with rational weight creates a completely different pattern.

These represent fundamentally different theories of persuasion that predict success in different psychological and competitive contexts.

When campaigns share message strategy dimensions, insights about what messaging angles work, what proof points resonate, what emotional tones convert—these insights transfer across wildly different product categories and target industries.

That's dimensional thinking in action.

Dimension 4: Execution Approach

The fourth dimension captures how campaigns are actually delivered—the structural choices about channels, sequences, personalization, and coordination that shape what's operationally possible and what resources are required.

This dimension often gets confused with "which channels performed well," but that's missing the point entirely. The dimension isn't about channel performance—it's about campaign architecture complexity and orchestration sophistication.

Channel Architecture

Some campaigns use single channels in isolation—an email campaign that just sends emails, a LinkedIn campaign that just runs ads. Others use multiple channels in coordinated sequences—email triggers LinkedIn ads for non-openers, webinar attendees receive personalized follow-up sequences, website visitors see coordinated display and social ads.

The architectural complexity of channel orchestration creates distinct campaign types that require different capabilities, different tools, and different expertise to execute well.

Simple single-channel campaigns with linear flows rhyme with other simple campaigns regardless of which specific channel, because the strategic pattern of focused execution and straightforward measurement transfers. A well-executed single-channel email nurture rhymes with a well-executed single-channel LinkedIn ad campaign in terms of operational complexity and optimization approaches.

Complex multi-channel campaigns with orchestrated sequences rhyme with other orchestrated campaigns—because the strategic pattern of timing coordination, message progression, and cross-channel measurement transfers. An account-based campaign using email, display, LinkedIn, and retargeting in coordinated sequences shares more operational DNA with another campaign using different channel combinations than it does with a simple email blast in the same industry.

Traditional benchmarks can't distinguish between these architectural patterns. They compare "email campaign performance" without recognizing that a simple broadcast and a sophisticated multi-touch nurture sequence are fundamentally different campaign types that should rhyme with different reference patterns.

Personalization Depth

Some campaigns treat audiences as homogeneous segments receiving identical messages. Others create dynamic experiences personalized by dozens of variables—company attributes, behavioral signals, engagement history, lifecycle stage, account context.

The sophistication of personalization creates distinct campaign types requiring different data infrastructure, different content operations, and different analytical capabilities.

Basic segmentation campaigns (treating all "enterprise chief information officers" the same) rhyme with other basic segmentation approaches, regardless of the channel, because the strategic pattern of segment-level messaging transfers. The operational challenges of managing 5–10 segments are similar whether you're doing it in email, display, or content syndication.

Advanced personalization campaigns (dynamically assembling messages based on 20+ variables per recipient) rhyme with other advanced personalization campaigns, regardless of the channel, because the strategic pattern of variable-driven content assembly transfers. The operational challenges of managing dynamic content at scale are similar whether you're doing it through email personalization engines, website customization, or programmatic display.

A moderately personalized email campaign rhymes more with a moderately personalized landing page experience than with a highly dynamic email campaign, because the operational complexity and content management challenges are more similar across channels at the same personalization tier than across personalization tiers in the same channel.

Nurture Sophistication

Some campaigns are single touchpoints—one email, one ad impression, one piece of content. Others are extended sequences—eight-email nurtures, 90-day account-based programs, multi-stage conversion funnels with branching logic.

The temporal complexity and sequence sophistication create distinct campaign patterns requiring different planning capabilities, different content operations, and different levels of patience from stakeholders expecting quick results.

Simple one-touch campaigns rhyme with other one-touch campaigns—whether webinar invitations, content downloads, or product announcements—because the strategic pattern of concentrated execution and immediate measurement transfers.

Complex multi-stage nurtures rhyme with other sophisticated sequences—whether email automation, account-based programs, or lifecycle marketing—because the strategic pattern of cumulative persuasion, timing optimization, and sequence testing transfers.

A 12-week email nurture campaign with branching logic based on engagement shares more operational DNA with a 12-week account-based program using multiple channels than with a single promotional email, even though both use email as a primary channel.

Content Volume

Some campaigns use minimal creative variations—one message, one design, one offer. Others require extensive creative production—dozens of message variations, multiple design treatments, A/B testing frameworks, iterative optimization.

The content production complexity creates distinct operational patterns requiring different creative resources, different approval processes, and different testing methodologies.

Low-variation campaigns (one message to all) rhyme with other low-variation approaches, regardless of the channel, because the strategic pattern of perfecting a single execution transfers. The creative development process is similar whether you're crafting one perfect email or one perfect landing page.

High-variation campaigns (dozens of message tests) rhyme with other high-variation approaches, because the strategic pattern of systematic testing and iterative optimization transfers. The creative operations challenges are similar whether you're testing email subject lines, creative ads, or landing page variations.

An email campaign testing 20 subject line variations shares more operational DNA with a display campaign testing 20 creative variations than with a single-message email campaign, because the testing infrastructure, approval workflows, and optimization methodologies are more similar across channels at the same variation complexity.

The Operational Implications

Execution complexity dimensions interact to create distinct operational profiles.

A simple, single-channel, low-personalization, one-touch campaign creates one operational pattern requiring specific capabilities. A complex, multi-channel, highly personalized, extended-nurture campaign creates a completely different operational pattern requiring entirely different capabilities.

Two campaigns with identical strategic goals and target audiences can have completely different execution complexity profiles—requiring different agency partnerships, different technology stacks, different team structures, and different budget allocations.

When campaigns share execution complexity dimensions, operational insights transfer—how to structure workflows, what tools to use, how to manage stakeholders, what problems to anticipate, and how long things actually take.

Traditional benchmarks ignore this dimension entirely, comparing campaign performance without accounting for things like when you're comparing a simple one-email promotion to a 90-day multi-channel orchestration program.

Dimensional pattern recognition recognizes that operational complexity predicts not just "what worked" but "what's actually replicable given your capabilities"—which is often the more important question.

Dimension 5: Business Goals

The fifth dimension captures what you're actually trying to achieve and how you're measuring success—the business context that determines whether a campaign "worked" and what optimization strategies make sense.

This dimension is critical because campaigns with identical performance metrics can be successes or failures depending on the business goals driving them.

Revenue Objectives

Some campaigns optimize for immediate revenue—closed deals, contract signatures, recognized revenue. Others optimize for pipeline development—qualified opportunities, sales conversations, future revenue potential.

This fundamental difference in time horizon and success criteria shapes everything about campaign strategy and performance evaluation.

Revenue-focused campaigns targeting late-stage buyers with near-term purchase intent require different strategies than pipeline-focused campaigns targeting early-stage buyers with 6–12-month evaluation cycles. The first optimizes for conversion rate and deal velocity. The second optimizes for engagement depth and relationship development.

A sales promotion campaign driving immediate purchases rhymes with other revenue-now campaigns regardless of product, because the strategic pattern of urgency creation, objection removal, and friction reduction transfers. The success metrics are similar (conversion rate, revenue per campaign, payback period). The optimization approaches are similar (testing offer structures, streamlining purchase paths, removing conversion barriers).

A thought leadership campaign building a future pipeline rhymes with other pipeline-development campaigns, because the strategic pattern of relationship cultivation, trust building, and long-term positioning transfers. The success metrics are similar (engagement depth, sales-accepted opportunities, pipeline value). The optimization approaches are similar (content quality, audience relevance, relationship progression).

Traditional benchmarks compare these fundamentally different campaign types as if they're equivalent because they use the same channel or target the same industry—completely missing that they're solving different business problems with different success criteria.

Conversion Priority

Some campaigns prioritize lead volume—maximizing the number of responses, downloads, or registrations regardless of qualification level. Others prioritize lead quality—generating fewer responses but ensuring they meet strict qualification criteria.

This strategic choice predicts campaign performance patterns and optimization approaches more reliably than most other variables.

Volume-focused campaigns with broad targeting and low-friction conversions rhyme with other volume campaigns regardless of channel, because the strategic pattern of maximizing top-of-funnel quantity transfers. The creative approach emphasizes broad appeal. The targeting prioritizes reach over precision. The conversion design minimizes barriers. The success metrics focus on cost-per-lead and total volume.

Quality-focused campaigns with narrow targeting and high-friction qualification rhyme with other quality campaigns, because the strategic pattern of attracting right-fit buyers transfers. The creative approach emphasizes specific value propositions. The conversion design includes qualification steps. The success metrics focus on sales-accepted rate and pipeline value per lead.

A content syndication campaign optimizing for maximum downloads shares more strategic DNA with a webinar campaign optimizing for maximum registrations than with a highly qualified buyer campaign in the same product category—because the volume-over-quality strategic choice creates similar optimization approaches and similar performance patterns.

Success Timeframe

Some campaigns measure success within days or weeks—immediate response, quick conversion, short feedback cycles. Others measure success across months or quarters—long attribution windows, complex multi-touch journeys, delayed conversion.

This temporal dimension shapes what's measurable, what's optimizable, and what insights you can extract from performance data.

Short-cycle campaigns with clear attribution rhyme with other short-cycle campaigns, because the strategic pattern of rapid testing and quick optimization transfers. You can run meaningful tests in weeks, iterate on creative quickly, and make data-driven decisions with confidence.

Long-cycle campaigns with complex attribution rhyme with other long-cycle campaigns, because the strategic pattern of patient optimization and proxy metrics transfers. You make strategic bets based on leading indicators, optimize intermediate engagement metrics, and accept that true success measurement takes quarters, not weeks.

A promotional email campaign with 48-hour conversion windows shares more measurement DNA with a limited-time offer display campaign than with a 90-day nurture campaign, because the attribution clarity and optimization speed are more similar across channels at the same time horizon than across time horizons in the same channel.

Business Context

Some campaigns operate in growth-mode contexts where the priority is market expansion, brand building, and customer acquisition, regardless of efficiency. Others operate in efficiency-mode contexts where the priority is cost optimization, ROI maximization, and doing more with less.

This business context profoundly shapes acceptable performance thresholds and optimization priorities.

Growth-mode campaigns can accept higher customer acquisition costs, longer payback periods, and broader targeting in pursuit of market share and brand awareness.

The success criteria emphasize market penetration and competitive positioning over immediate profitability.

Efficiency-mode campaigns require tight cost controls, quick payback, and precisely targeted spending. The success criteria emphasize return on investment and cost per acquisition over market share expansion.

A well-funded startup's customer acquisition campaign shares strategic patterns with other growth-mode campaigns, regardless of industry, because the business context of "grow fast, optimize later" creates similar strategic choices. High spending, broad experimentation, and market share focus transfer across contexts.

A mature company's cost-conscious campaign shares strategic patterns with other efficiency-mode campaigns—because the business context of "prove ROI, minimize waste" creates similar constraints. Careful targeting, measurement rigor, and efficiency optimization transfer.

Metric Sophistication

Some organizations measure campaigns with simple metrics—opens, clicks, downloads, registrations. Others employ sophisticated measurement frameworks—multi-touch attribution, incrementality testing, cohort analysis, predictive modeling.

This measurement sophistication creates distinct campaign types requiring different analytical capabilities and different optimization approaches.

Simply measured campaigns rhyme with other simply measured campaigns, because the strategic pattern of optimizing obvious metrics transfers. You improve open rates, increase click-through rates, and boost conversion rates using straightforward A/B testing.

Sophisticatedly measured campaigns rhyme with other sophisticated-measurement contexts, because the strategic pattern of multi-variable optimization and statistical rigor transfers. You balance multiple objectives across attribution models, test incrementality, and optimize for long-term value.

A campaign measured purely by lead volume shares more optimization DNA with another volume-measured campaign than with a campaign using complex attribution modeling in the same channel, because the measurement sophistication determines what's actually optimizable.

The Business Context Combinations

Business goal dimensions interact to create distinct success frameworks.

A revenue-focused, quality-prioritized, short-cycle, efficiency-mode campaign creates one success pattern. A pipeline-focused, volume-prioritized, long-cycle, growth-mode campaign creates a completely different success pattern.

These represent fundamentally different business contexts that make campaigns with identical execution look like successes or failures depending on the goals driving them.

When campaigns share business goal dimensions, performance insights transfer—not just "what metrics were achieved" but "what those metrics meant" and "whether the campaign succeeded given the business context."

Traditional benchmarks completely ignore this dimension. They report "average performance" without any framework for understanding whether those averages represent successful campaigns, failed campaigns, or meaningless noise from mismatched business contexts.

Dimensional pattern recognition maps business goal contexts and identifies campaigns solving similar business problems with similar success criteria—making performance comparisons actually meaningful instead of statistically meaningless noise.

Seeing the Five Dimensions in Practice

Let me show you what dimensional mapping actually looks like with a real campaign, so you can see how these abstract concepts translate into strategic intelligence.

> **The campaign:** A digital transformation consultancy launching a multi-channel campaign targeting chief operating officers at traditional retail chains about modernizing their supply chain operations.

Dimension 1—Audience behavior:

- Buyer sophistication: Moderate (understand business need but not technical details)

- Decision authority: High (C-suite with budget control but needs CEO alignment)

- Risk tolerance: Low (transformation failures are highly visible)

- Awareness stage: Problem-aware (know they're falling behind digitally, unsure of solution path)

- Psychological state: Urgency mixed with paralysis (know they must change, but overwhelmed by options)

Dimension 2—Competitive environment:

- Market maturity: Emerging but accelerating (post-COVID digital acceleration)

- Competitive intensity: Moderate (big consultancies dominate but leave gaps)

- Positioning: Specialist alternative (boutique firm vs. Accenture/McKinsey)

- Budget environment: Available but scrutinized (transformation is funded but must show quick wins)

- Timing context: Compressed (Amazon pressure making speed critical)

Dimension 3—Message strategy:

- Core promise: Managed transformation ("we've done this 100 times," not "innovative approaches")

- Positioning approach: Experience-driven (case studies from similar retailers)

- Objection strategy: Reframe ("the risk isn't change, it's standing still")

- Emotional weight: 60% emotional/40% rational (fear of irrelevance drives decisions)

- Sophistication level: Business-focused (ROI and timelines over technical architecture)

Dimension 4—Execution approach:
- Channel architecture: Account-based (personalized outreach to 50 target companies)

- Personalization depth: High (custom analysis of each retailer's digital gaps)

- Nurture sophistication: Short burst (three-week executive engagement campaign)

- Content volume: Moderate (five core pieces with company-specific customization)

- Orchestration complexity: Moderate (LinkedIn, email, executive events)

Dimension 5—Business goals:
- Revenue objective: Large deal capture (targeting $1M+ engagements)

- Conversion priority: Quality over quantity (five executive conversations worth 500 leads)

- Success timeframe: Medium-cycle (3–4 month decision process)

- Business context: Growth mode (investing to establish market position)

THE FIVE DIMENSIONS

- Metric sophistication: Simple (meetings booked with target executives)

Now here's where pattern recognition becomes powerful:

This campaign's dimensional signature reveals it will likely rhyme with campaigns from completely unexpected places.

For instance, a B2C luxury fitness brand targeting affluent suburban families for new membership might share striking dimensional similarity:

- Same moderate sophistication with high purchase authority (Dimension 1: Audience behavior)

- Same accelerating market with specialist positioning (Dimension 2: Competitive environment)

- Same experience-driven, emotionally weighted messaging (Dimension 3: Message strategy)

- Different channels but similar account-based personalization (Dimension 4: Execution approach)

- Same quality-over-quantity with medium-cycle conversion (Dimension 5: Business goals)

Or consider an employee readiness platform targeting CHROs at Fortune 500 companies about workforce adaptation:

- Same urgency-mixed-with-paralysis psychology (Dimension 1: Audience behavior)

- Same emerging market with boutique-vs.-giant dynamics (Dimension 2: Competitive environment)

- Same "we've done this before" managed-change messaging (Dimension 3: Message strategy)

- Similar executive-focused orchestration (Dimension 4: Execution approach)

- Similar large-deal, quality-focused goals (Dimension 5: Business goals)

The dimensional signatures suggest all three campaigns would benefit from similar strategic approaches: leading with transformation expertise, not innovation; personalized executive engagement over broad demand generation; emotional urgency balanced with rational proof; and quick wins to build confidence.

Meanwhile, another digital transformation campaign targeting the same COOs but with different dimensional characteristics—say, a pure technology play targeting tech-savvy early adopters with self-service tools—would have almost nothing in common strategically despite being in the same category.

The dimensional signature predicts strategic success patterns far better than surface categories.

How Dimensions Interact: The Reality of Pattern Recognition

Here's what makes dimensional thinking so powerful and so complex: these five dimensions don't operate independently. They interact, combine, and create emergent patterns that predict campaign performance in ways no single dimension can explain.

A campaign's dimensional signature is the combination of where it sits across all five dimensions simultaneously.

Consider two campaigns:

Campaign A: Targets risk-averse, experienced buyers with decision authority (audience behavior) in a mature, competitive market fighting incumbent inertia (competitive environment) using loss-avoidance, proof-driven messaging (message strategy) with complex multi-channel orchestration (execution approach) optimizing for a quality pipeline with long attribution windows (business goals).

Campaign B: Targets risk-tolerant, first-time buyers with influence, not authority (audience behavior), in an emerging category with minimal competition (competitive environment) using gain-focused, thought-leadership messaging (message strategy) with simple single-channel execution (execution approach) optimizing for lead volume with short conversion cycles (business goals).

These campaigns share nothing dimensionally. Every strategic choice that works for Campaign A likely fails for Campaign B, and vice versa.

They could both be "B2B SaaS email campaigns targeting mid-market buyers"—identical in traditional benchmark categories—yet have zero transferable insights because their dimensional signatures are completely different.

Now consider two more campaigns:

Campaign C: A cybersecurity campaign targeting risk-averse CISOs at financial institutions in a competitive market (audience behavior & competitive environment) using fear-reduction, proof-driven messaging (message strategy) with moderate complexity (execution approach), optimizing for qualified opportunities (business goals).

Campaign D: A compliance software campaign targeting risk-averse general counsels at healthcare providers in a competitive market (audience behavior & competitive environment) using regulatory-risk, proof-driven messaging (message strategy) with moderate complexity (execution approach), optimizing for qualified opportunities (business goals).

These campaigns share dimensional signatures despite being completely different industries, products, and target personas.

The strategic insights transfer. The psychological patterns rhyme. The optimization approaches work similarly.

That's what Chapter 5 demonstrated through scenarios. This is the framework explaining why those scenarios worked.

Dimensional Weighting: Not All Dimensions Matter Equally

Here's another level of sophistication: not all dimensions matter equally for every strategic question you're trying to answer.

If you're asking, "what message strategy should I use?", Dimensions 1, 2, and 3 (audience behavior, competitive environment, message strategy) matter enormously. Dimensions 4 and 5 (execution approach, business goals) matter less for that specific question.

If you're asking, "how should I structure my campaign workflow?", Dimension 4 (execution approach) matters most. The others provide useful context but aren't primary predictors.

If you're asking, "how should I measure success?", Dimension 5 (business goals) dominates. The others help interpret metrics, but don't define what success means.

This context-dependent weighting is where the human-AI partnership becomes essential. AI can calculate similarity scores across all dimensions. But knowing which dimensional similarities matter most for your specific strategic question requires the kind of expertise that comes from leading campaigns across diverse contexts and recognizing which patterns actually predict outcomes.

This is why sophisticated AI orchestration alone isn't enough. You need experienced marketing professionals who can frame the right questions, weight the relevant dimensions, and interpret what mathematical patterns mean strategically. The AI amplifies human expertise—it doesn't replace it.

You can't just calculate similarity scores across all five dimensions and assume that campaigns with the highest overall similarity scores are the most comparable. Sometimes, a strong match on Dimensions 1 and 3 with mismatches on others creates valuable insights. Sometimes, Dimension 2 similarity matters more than anything else.

Pattern recognition isn't just mapping campaigns across dimensions. It's knowing which dimensional patterns predict which strategic insights for which questions—and that requires both computational power and sophisticated judgment working together.

Not All Rhymes Are Equally Relevant

Here's a critical insight that separates effective pattern recognition from noise: finding campaigns that rhyme dimensionally is step one. Knowing which dimensional rhymes matter for your specific strategic question is step two.

A campaign rhyming across Dimensions 1 and 3 might offer profound message strategy insights while teaching you nothing about operational execution. A campaign rhyming on Dimension 4 might show you how to structure workflows while having zero transferable positioning insights. A campaign rhyming on Dimension 5 might validate your success metrics while saying nothing about the best creative approach.

This isn't a limitation—it's precision.

You're not looking for campaigns that rhyme across all dimensions. You're looking for campaigns that rhyme across the dimensions that matter for the specific question you're trying to answer.

Which dimensional rhymes are relevant depends entirely on context:

- Planning a message strategy? Dimensions 1, 2, and 3 matter most.

- Structuring operations? Dimension 4 dominates.

- Setting success criteria? Dimension 5 is primary.

- Evaluating a total campaign approach? All five matter, but they should be weighted differently.

This is where the human-AI partnership becomes particularly valuable. AI can identify all campaigns sharing dimensional similarities with yours. (We'll cover how to do this in Chapter 8.) But experienced marketing judgment determines which of those rhymes will actually inform your specific decision. The computational power finds the patterns. The strategic expert knows what to do with them.

Without this filtering, you'd drown in dimensional matches that are mathematically valid but strategically irrelevant to your question. With it, you get targeted insights from the campaigns that rhyme in ways that matter for what you're trying to accomplish.

That's why this isn't a DIY framework. Understanding the dimensions is a great start; in the rest of the book, we'll cover how to apply them effectively at scale.

The Curse of Dimensionality: Why This Gets Complicated Fast

Here's the mathematical reality: when you map campaigns across five dimension categories, each with multiple sub-dimensions, you're working in extremely high-dimensional space.

Dimension 1 (audience behavior) has at least 5 meaningful sub-dimensions. Dimension 2 (competitive environment) has another 4–5. Dimensions 3, 4, and 5 each add several more.

You're not comparing campaigns across five variables. You're comparing campaigns across 20–30+ variables simultaneously.

In high-dimensional spaces, several counterintuitive things happen:

Most campaigns are far apart

When you're comparing across dozens of dimensions, most campaigns share very little dimensional similarity. The number of truly comparable campaigns shrinks dramatically as dimensional sophistication increases. This is actually good—it means you're being more precise about what "comparable" means—but it requires access to thousands of campaigns to find enough dimensional matches to extract reliable patterns.

Distance metrics get weird

In high-dimensional spaces, all pairwise distances start looking similar. The difference between "very similar" and "somewhat similar" compresses. Determining meaningful similarity thresholds requires sophisticated statistical approaches and extensive empirical validation.

Irrelevant dimensions add noise

Including dimensions that don't actually predict the outcomes you care about makes pattern recognition less effective, not more. Knowing which dimensions matter for which strategic questions requires both mathematical sophistication and domain expertise.

Interaction effects dominate

The way dimensions combine matters more than individual dimensional values. Two campaigns matching on Dimensions 1 and 3 might be very comparable if they also match on Dimension 2, but not comparable at all if Dimension 2 differs. These interaction effects require analyzing thousands of campaign combinations to recognize reliable patterns.

This is the computational and analytical challenge that makes dimensional pattern recognition simultaneously more valuable than benchmarks and harder to execute than it appears.

Anyone can understand the five dimensions conceptually. Actually mapping campaigns accurately, calculating meaningful similarity, and extracting reliable insights requires sophisticated AI infrastructure working in concert with deep marketing expertise.

The AI handles computational challenges humans can't manage: processing thousands of campaigns across dozens of variables, calculating complex similarity metrics, and identifying non-obvious patterns. The human expert handles interpretive challenges AI can't manage: knowing which patterns actually predict outcomes, understanding contextual nuances that don't reduce to variables, and recognizing when mathematical correlations don't represent strategic causation.

Neither works alone. Together, they make dimensional pattern recognition possible at scale.

Why Category Thinking Fails: The Dimensional Explanation

Traditional benchmarks fail because they use low-dimensional category thinking to organize high-dimensional reality.

"B2B SaaS email campaigns" groups campaigns by three variables: business model, product type, and channel. These are real dimensions—but they're not the dimensions that predict strategic performance patterns.

Two campaigns can match perfectly on business model, product type, and channel while differing across all five meaningful dimensions—making them completely non-comparable for strategic purposes.

Conversely, two campaigns can differ completely on business model, product type, and channel while matching across all five meaningful dimensions—making them highly comparable for strategic insights.

Traditional categories are organizational conveniences. They make reporting easier, stakeholder communication simpler, and benchmark publishing possible.

But they're not predictive. They don't map to the dimensions that actually matter.

Dimensional pattern recognition is harder. It requires more sophisticated thinking, more complex analytical infrastructure, and more extensive campaign data.

But it actually works. It finds genuinely comparable campaigns and surfaces insights that transfer.

That's the tradeoff: comfortable simplicity that misleads versus sophisticated complexity that illuminates.

What This Means for You

Let's bring this back to practical reality.

You now understand the five dimensions that predict whether strategic insights transfer between campaigns:

1. Audience behavior

2. Competitive environment

3. Message strategy

4. Execution approach

5. Business goals

You understand that these dimensions interact, that weighting matters, and that high-dimensional pattern recognition is computationally complex.

So what do you do with this framework?

If you're a marketing leader evaluating campaigns, stop asking "what's the benchmark for campaigns like mine?" Start asking "what campaigns share dimensional signatures with mine, and what did they learn?"

You won't be able to map campaigns across all dimensions with mathematical precision—but you can start thinking dimensionally about what makes campaigns truly comparable.

When someone shows you a "successful B2B email campaign," ask questions like these:

- What was the audience psychology? (experienced buyers or first-time? risk-averse or risk-tolerant?)

- What was the competitive context? (mature category or emerging? crowded market or open field?)

- What was the message strategy? (gain-focused or fear-reduction? thought leadership or proof-driven?)

- What was the execution complexity? (simple broadcast or orchestrated sequence?)

- What were the business goals? (immediate revenue or long-term pipeline?)

If those dimensional answers don't match your context, that campaign's performance metrics tell you nothing useful about what you should expect.

If you're an agency evaluating cross-client insights, dimensional thinking changes how you leverage your portfolio experience.

Stop organizing insights by client industry. Start organizing insights by dimensional patterns.

Your cybersecurity client's campaign might teach your compliance client more than another cybersecurity campaign taught them—if the dimensional signatures match.

Your client roster isn't a collection of isolated accounts. It's a portfolio of campaigns spanning different dimensional combinations. The dimensional overlaps are where transferable insights hide.

If you're a strategist planning campaigns, dimensional thinking changes how you identify relevant case studies and comparable examples.

Stop looking for campaigns in your industry or using your channel. Start looking for campaigns sharing dimensional characteristics.

If you're planning a campaign targeting sophisticated, risk-averse buyers in a competitive market with proof-driven messaging, find campaigns matching those dimensions—regardless of industry, product, or even channel. Those are the campaigns with transferable strategic insights.

If you recognize the value but lack the capability, you've discovered exactly why this approach requires both sophisticated AI orchestration and deep marketing expertise working together.

Understanding dimensional thinking doesn't mean you can replicate it at scale. Just like understanding flight physics doesn't mean you can build airplanes.

The gap between framework comprehension and reliable implementation is where the human-AI partnership matters most. You need computational power to process the complexity and experienced judgment to interpret what it means. You need AI to find the patterns and marketing expertise to know which patterns are signal versus noise.

That's not a limitation—it's the reality of sophisticated intelligence work in complex domains.

The Bridge to What's Next

This chapter gave you the dimensional framework—what the five core dimensions are, why they matter, how they interact, and why high-dimensional pattern recognition is both more valuable and more complex than traditional benchmarking.

But I haven't explained how to actually find dimensional patterns at scale. I haven't shown you how to map thousands of campaigns, calculate meaningful similarity, or extract reliable insights from high-dimensional spaces.

That's intentional.

Because the next critical question isn't "how do you map campaigns mathematically?"

It's "why does finding these patterns require both AI capabilities *and* deep human expertise—and why will that remain true even as AI gets more powerful?"

THE FIVE DIMENSIONS

Chapter 7 explores the human-AI partnership required for dimensional pattern recognition at scale. It's the strategic core of this entire book—the explanation of why this approach creates a defensible, sustainable competitive advantage that actually appreciates over time rather than depreciating with each new AI model release.

Then, Chapter 8 will show you how AI enables pattern recognition that was previously impossible—the technical capabilities that make dimensional thinking actionable at scale.

But before we could get to the technology, we needed to understand the human expertise that makes the technology valuable.

Because here's what 25+ years of leading campaigns has taught me: the hard part isn't calculating similarity scores. The hard part is knowing what those scores mean, which patterns actually matter, and how to translate mathematical correlations into strategic insights that win in the market.

That's where we're going next.

But first, sit with what you've learned about dimensions.

Look at your current campaigns through this lens. Think about past successes and failures dimensionally.

Can you articulate where your campaigns sit across these five dimensions?

Can you identify other campaigns—possibly in completely different industries or channels—that share dimensional signatures?

Can you see why traditional benchmarks that ignore these dimensions are comparing incomparable campaigns?

If you can, you're starting to see the pattern recognition that transforms campaign performance.

If you can't—or if you can see it conceptually but recognize you'd need thousands of mapped campaigns and sophisticated analytical infrastructure to extract reliable insights—you're understanding exactly why this approach requires specialized expertise to execute well.

Either way, you now know what dimensions matter.

And that changes everything about how you think about campaign comparison.

Chapter Seven

Finding Rhymes—The Human-AI Moat

MOST PEOPLE ASSUME THAT as AI gets better, human expertise becomes less valuable. Better AI tools → less need for skilled practitioners → commodity work → compressed margins.

That's the *Terminator* premise: machines inevitably replace humans as capabilities advance.

But here's what's actually happening in sophisticated strategic work: *AI advancement makes experienced human expertise* more *valuable, not less.*

This isn't wishful thinking or clever positioning. It's an economic reality emerging from how AI capabilities interact with complex domains like marketing campaign intelligence. And understanding why this happens—why the "moat" around human expertise actually widens as AI advances—is essential for anyone building a sustainable competitive advantage in dimensional pattern recognition.

AI can now generate hundreds of campaign variations in minutes, analyze massive datasets instantly, and produce professional-quality content that would have taken teams weeks to create five years ago. All of that is true.

And yet the marketers best positioned to capitalize on these capabilities aren't necessarily the ones who adopted AI earliest or who understand the technology most deeply. They're the ones with decades of cross-industry campaign experience, multi-disciplinary strategic judgment, and institutional memory about what actually works in diverse contexts.

The technology enables scale. The expertise determines value.

As AI commoditizes tactical execution, strategic judgment becomes the scarce resource. And that judgment—knowing which patterns matter, understanding how context modifies pattern applicability, recognizing when sophisticated AI outputs are strategically naive—is built through years of exposure to diverse campaign contexts and cannot be shortcut.

This creates a counterintuitive moat that strengthens as AI advances.

Let me show you how this works.

The Three Phases of AI Value

To understand why human expertise becomes more valuable as AI advances, we need to understand how AI creates different types of value at different levels of sophistication.

Phase 1: Efficiency Gains

This is where most AI adoption starts. You use AI to do things you were already doing, just faster:

- Generate email subject line variations in seconds instead of hours.
- Draft social media content quickly.
- Produce first-draft copy that humans then polish.
- Summarize meeting notes automatically.
- Schedule optimizations based on performance data.

These are real productivity improvements. Tasks that took hours now take minutes. Teams can produce more output with the same headcount.

But *everyone gets these efficiency gains simultaneously*. When AI tools for content generation become widely available, every agency and brand has access. The playing field levels within 6–12 months of any new capability launch.

There's no sustainable competitive advantage in Phase 1 AI adoption because efficiency tools commoditize rapidly. Your competitor, using the same AI tools, gets the same speed improvements you do.

This is valuable for operational efficiency, but it's not a moat. It's table stakes.

Phase 2: Scale That Previously Wasn't Feasible

The next level is using AI to do things that weren't practically possible before:

- Generating 500 creative variations for multi-variate testing programs

- Analyzing campaign performance across dozens of dimensional attributes simultaneously

- Processing thousands of campaigns to surface pattern candidates

- Personalizing content at scale beyond what manual processes could handle

- Testing strategic hypotheses that would require too much manual work to explore systematically

This is where AI starts creating genuine new capabilities rather than just accelerating existing workflows. You're not just doing the same work faster—you're doing work that wouldn't happen at all without the computational power of AI.

But there's a subtle trap: *scale without strategic judgment creates noise, not insight.*

Generating 500 creative variations only creates value if you can evaluate which variations actually align with strategic goals versus which just sound good algorithmically. Analyzing campaigns across dozens of dimensions only matters if you can interpret what those dimensional patterns mean for your specific context. Testing more hypotheses faster just burns budget if you're testing low-probability approaches.

AI gives you scale. Human expertise gives you judgment about what to do with that scale.

Without experienced judgment informed by pattern recognition across hundreds of real campaigns in diverse contexts, AI-enabled scale becomes overwhelming rather than advantageous. You either pick randomly, follow whatever AI suggests most confidently (which often leads to expensive mistakes), or spend more time evaluating options than you save by having AI generate them.

Phase 3: Sophistication That Requires Human-AI Partnership

The third level—where sustainable competitive advantage emerges—is using AI to enable strategic work that requires both computational power *and* deep human expertise:

- Dimensional pattern recognition across thousands of campaigns

- Strategic hypothesis generation informed by multi-domain marketing experience

- Quality control on AI outputs that improves as AI gets better (counterintuitively)

- Orchestration of multiple AI capabilities for different analytical tasks

- Interpretation of what mathematical patterns mean strategically

This is where human expertise and AI capability become genuinely complementary rather than substitutable.

The AI provides computational scale—comparing your campaign across dimensions to thousands of others, surfacing similarity patterns, and generating hypotheses.

The human expertise provides strategic judgment—determining which dimensional patterns actually matter for your context, validating that calculated similarities represent genuine strategic opportunities rather than spurious correlations, interpreting what patterns mean, and knowing when to trust patterns versus when context overrides.

Neither works alone when it comes to sophistication.

AI without human expertise produces plausible-sounding but strategically naive suggestions. It finds patterns that don't transfer. It optimizes for metrics that don't drive outcomes. It makes recommendations that work in historical data but fail when market conditions shift.

Human expertise without AI hits cognitive limits—you can't manually analyze hundreds of campaigns across dozens of dimensions simultaneously. You can't systematically surface non-obvious rhymes that only become visible in high-dimensional pattern space. You can't validate intuition rigorously across a sufficient scale.

But combine deep marketing expertise with sophisticated AI orchestration, and you create capabilities that neither achieves alone—and that compound over time as AI tools advance and human pattern recognition expertise accumulates.

This is where the defensible moat forms.

Because the sophistication ceiling—where strategic insight matters more than operational speed—is where expertise determines value. And unlike technology access, which equalizes quickly, expertise development takes years and continues compounding.

The better AI gets at execution, the more each strategic decision matters.

The more options AI generates, the more valuable judgment becomes.

The more sophisticated AI outputs look, the harder it becomes to distinguish genuinely strategic recommendations from plausible-but-wrong suggestions.

This is the paradox: AI advancement increases the leverage of expertise; it doesn't decrease it.

Why Off-the-Shelf AI Tools Hit a Ceiling

To understand why sophisticated pattern recognition requires human expertise orchestrating AI—rather than just AI alone—look at what happens when companies try to solve this problem with generic tools.

The Generic AI SaaS Tool Approach

Pick any of the current AI SaaS marketing tools: Jasper, Copy.ai, Anyword, Writer, or any of the dozens of others. They all follow the same pattern:

- Provide a user interface for accessing a large language model (usually GPT-5).

- Add some marketing-specific prompt templates.

- Include basic features like tone control, brand voice settings, and content variation generation.

- Charge a monthly subscription.

These tools are useful for basic content generation. If you need 20 email subject line variations or want to quickly draft social media posts, they work fine.

But they hit a hard ceiling when you try to use them for sophisticated pattern recognition or strategic intelligence:

They use single AI models (or occasionally two).

This means you're limited by that model's specific strengths and blind spots. When GPT-5 struggles with a particular analytical task or hallucinates patterns that don't exist, you have no validation mechanism. You can't cross-check against other models with different training and architecture.

They have no performance intelligence.

They can generate content, but they can't tell you which dimensional patterns predict success because they've never analyzed thousands of campaigns across multiple contexts. They're creation tools, not intelligence systems.

They produce the same outputs for everyone.

Your competitor, using the same tool with similar prompts, gets similar suggestions. There's no proprietary insight, no compounding advantage, no moat. The tool is a commodity available to anyone willing to pay the subscription fee.

They can't validate their own suggestions.

When the AI proposes a campaign approach, these tools have no framework for evaluating whether that approach actually works. They generate plausible-sounding content, but "plausible-sounding" and "strategically sound" are very different things.

Within 6–12 months of launch, any advantage from these tools commoditizes completely. Everyone has access. Everyone can generate similar outputs. The playing field levels.

The Platform AI Problem

There's an even more insidious version of off-the-shelf AI that most marketers don't think critically about: the AI built into advertising platforms themselves.

Google Ads now has AI-powered campaign optimization. Meta has Advantage+ campaigns. LinkedIn has automated bidding and creative optimization. TikTok has Smart Performance campaigns. Every major platform has integrated AI deeply into its self-service interfaces.

On the surface, this seems convenient. The platform analyzes your performance data, suggests budget adjustments, recommends audience expansions, and even generates creative variations. Why would you need human expertise when the platform's AI handles everything?

I was at an invite-only dinner in Boston—one of those intimate industry evenings where real conversation happens—when Mike O'Toole, president of PJA Marketing and Advertising, made an observation that crystallized something I'd been sensing but hadn't articulated clearly:

"Google's AI wants you to spend more on Google Ads. Meta's AI wants you to increase Facebook budgets. Their 'optimization recommendations' are aligned with their shareholder value, not your campaign outcomes."

This wasn't cynicism—it was a recognition of basic economic reality: *platform AI optimizes for platform revenue, not your business outcomes.*

When Meta's AI suggests "increase budget by 40%" or "expand to more placements," is that because it's genuinely the best strategy for your business? Or because Meta profits from larger budgets and broader reach, regardless of whether that incremental spend delivers proportional results for you?

When Google's Performance Max recommends moving budget from search into display and YouTube, is that driven by what performs best for your specific goals? Or by Google's strategic imperative to grow revenue in its less-saturated inventory?

These platforms are sophisticated. Their AI is genuinely advanced. But it's fundamentally programmed to maximize platform revenue. The AI might be optimizing beautifully—just not for what you care about.

This is a structural conflict of interest that no amount of AI sophistication overcomes. An objective intelligence system—one with no financial stake in where you allocate media budget—can honestly say "you're overspending on Meta and should shift budget to LinkedIn" or "this channel isn't working for your audience profile."

Platform AI will never tell you to spend less on their platform. It will never suggest you'd be better off elsewhere. It's literally designed to keep you spending more, not to optimize your total marketing efficiency.

And you can't audit it. You can't request "show me why you're recommending this increase" and get a transparent dimensional analysis. The recommendations come from black box algorithms trained on billions of advertisers, optimized for aggregate platform revenue, not your specific strategic context.

The same marketers who would never let a media agency mark up their spend by 20% will blindly follow platform AI recommendations that effectively do the same thing—just more opaquely.

The Black Box System Approach

Some companies try to solve the AI marketing challenge by building proprietary "AI marketing systems" that make decisions automatically. These typically work like this:

- Feed campaign data into a machine learning system.

- Let algorithms optimize targeting, messaging, and budget allocation.

- Trust the AI to figure out what works without human intervention.

The problem with black box systems is that you can't interrogate them. When they suggest a campaign approach or optimization, you can't ask "why?" You can't validate whether the pattern the AI detected is a real signal or a spurious correlation. You can't apply your judgment about whether the suggestion makes strategic sense, given competitive context, audience psychology, or timing dynamics.

This is dangerous for strategic decisions. AI can spot mathematical patterns in data that don't represent true causal relationships. It can optimize for metrics that don't actually drive business outcomes. It can make suggestions that work in narrow historical contexts but fail when market conditions shift.

Without the ability to understand, validate, and override AI suggestions based on expert judgment, black box systems create brittle advantages that collapse when their underlying assumptions break.

The Human-Only Approach

The alternative extreme—experienced humans working without AI assistance—also hits scaling limits.

A skilled marketing strategist can hold maybe 15–20 campaigns in working memory and recognize dimensional patterns across them. They can recall examples from their own experience that rhyme with current challenges. They can apply intuitive pattern recognition built through years of exposure to what works.

But they can't manually analyze 500 campaigns across 60+ dimensions simultaneously. They can't calculate multi-dimensional similarity scores accounting for interaction effects and context-dependent weighting. They can't systematically surface non-obvious rhymes that only become visible in high-dimensional pattern space.

Human cognition is remarkable at pattern recognition within certain constraints. But it's not built for processing the scale and dimensionality required for sophisticated cross-campaign intelligence.

This is why the sophisticated approach requires orchestrated intelligence: experienced marketing experts working in partnership with AI capabilities, each amplifying what the other does well.

The Multi-LLM Orchestration Advantage

Here's where most people miss a critical strategic insight about AI: the competitive landscape among AI models isn't converging toward uniformity—it's diverging toward specialization.

In 2023, when there were only a handful of advanced models, people assumed they would become increasingly similar as they all improved. The thinking was: "Eventually, all models will be great at everything, so it won't matter which one you use."

That's not what's happening.

Competitive Pressure Drives Differentiation

As of early 2026, we have Claude Sonnet 4.5, GPT-5.2, Gemini 3.0, Llama 4, Mistral Large 3, Grok 4.1, Nova 2 and dozens of other models—with more launching constantly. Each is backed by companies competing intensely for market share and answering to shareholders with different strategic priorities.

That competitive pressure doesn't push them toward sameness. It pushes them toward differentiation. Each model optimizes for different strengths based on what their parent company's shareholders value.

As of today, Claude models excel at nuanced strategic reasoning, following complex instructions, and maintaining a consistent brand voice across long-form content. They're particularly strong at understanding context and applying judgment about what messaging approaches fit specific audience psychologies.

GPT-5 excels at creative variation generation, brainstorming diverse approaches, and rapidly producing high-quality copy across multiple formats. It has broad knowledge and generates more stylistic variety than other models.

Gemini models excel at analytical synthesis, processing large datasets, identifying patterns in performance data, and generating structured output. They're particularly strong at taking messy inputs and producing organized strategic frameworks.

Specialized models like Llama or Mistral might optimize for speed, cost-efficiency, or specific domain expertise.

But here's the critical point: *these relative strengths will shift as models evolve.*

Anthropic might prioritize speed over reasoning depth in their next release based on market feedback. OpenAI might optimize for cost reduction over creative quality to hit profitability targets. Google might push Gemini toward multimodal capabilities at the expense of pure text analysis.

The companies behind these models answer to shareholders and market dynamics, not to marketers' ideal use cases. Development priorities shift based on competitive positioning, revenue models, and strategic goals that have nothing to do with which capabilities matter most for sophisticated marketing intelligence.

This creates two critical insights.

First, *model differentiation will accelerate, not disappear.* As models improve, their relative strengths become more pronounced rather than converging. A model that's 10% better at creative ideation and 10% worse at data analysis might not show huge differences when both capabilities are mediocre. But when both capabilities are sophisticated, that 10% difference becomes strategically significant.

Second, *you need architectural flexibility to adapt as the landscape shifts.* The optimal model for strategic analysis today might not be optimal six months from now. Being locked into hard-coded integrations with specific models creates technical debt and limits your ability to leverage advances across the AI ecosystem.

The Orchestration Insight

If different models have different strengths—and those strengths will continue differentiating and shifting based on shareholder-driven priorities rather than user needs—then the sustainable competitive advantage isn't in having access to one great model. It's in knowing which model to use for which task, in what sequence, with what validation, and having the technical architecture to adapt as capabilities evolve.

Consider sophisticated pattern recognition for campaign intelligence:

Strategic analysis of dimensional similarities

Today, you might use Claude for its strong reasoning about why campaigns might rhyme across non-obvious dimensions, its ability to understand nuanced audience psychology and competitive dynamics, and its skill at explaining which patterns matter strategically versus which are just statistical noise. Tomorrow, if another model surpasses Claude's reasoning capabilities, you need the flexibility to shift.

Creative variation generation from pattern insights

Currently, GPT-5 excels at rapid ideation, generating diverse creative approaches informed by dimensional patterns, and producing high-quality copy across multiple formats and tones. But if Anthropic's next Claude release focuses on creative capabilities, or if a new model emerges with superior creative variation, you need architecture that lets you adapt.

Performance data synthesis and pattern validation

Gemini currently shows strength at processing campaign performance data, identifying which dimensional patterns actually correlate with better outcomes, and structuring insights into actionable frameworks. As Google's priorities shift, these relative advantages might change.

Cross-model validation

When multiple models independently identify the same dimensional pattern or strategic insight, your confidence in their output increases. When they disagree, that disagreement itself is informative—it highlights edge cases or contextual dependencies that require human judgment.

This isn't just "use the best tool for each job." It's building a system where model diversity creates emergent intelligence that no single model provides, while maintaining the architectural flexibility to adapt as the competitive landscape evolves.

As models get better and more differentiated, orchestration advantage compounds rather than eroding—but only if you've built model-agnostic architecture.

If you're locked into a single model—whether by vendor contract, technical architecture, or just habit—you're constrained by that model's specific limitations and vulnerable to that vendor's strategic decisions. When it fails at a task, you're stuck. When it optimizes for capabilities you don't need (because that's what shareholders want), you can't benefit from competitors' different optimization choices.

If you're orchestrating multiple models based on a deep understanding of their relative strengths, with technical architecture that makes swapping models low-friction, you benefit from every advancement across the entire AI landscape. When Claude releases a better reasoning model, you gain strategic analysis improvements. When GPT-6 launches with stronger creative capabilities, you gain content generation improvements. When Gemini improves its analytical processing, you gain pattern validation improvements.

Your competitive advantage appreciates continuously as the AI ecosystem evolves, rather than depending on any single vendor's roadmap or being held hostage to shareholder-driven development priorities that don't align with your needs.

But orchestration requires judgment about orchestration. You can't just "use all the models for everything." You need to understand which model to use when, how to structure prompts for each model's specific strengths, how to validate outputs across models, when to trust AI suggestions versus when to override them with human expertise, and when model capabilities have shifted enough to warrant changing your orchestration approach.

That judgment—knowing how to orchestrate AI effectively for sophisticated strategic work and adapt as the landscape evolves—is itself a form of expertise that compounds over time and doesn't commoditize.

Why Institutional Memory Compounds While AI Resets

Here's another critical distinction between human expertise and AI capability:

AI models reset with each new conversation. Human expertise accumulates and transfers across contexts.

When you start a new conversation with Claude or GPT-5 or any other model, it has no memory of previous conversations (unless you explicitly provide that context). Each interaction begins fresh. The model can't say "remember last week when we analyzed that campaign and discovered X?" It has no institutional memory.

Humans do. And sophisticated organizations augment human memory with technical architecture designed to persist and retrieve pattern intelligence.

Human Institutional Memory

An experienced marketing strategist who analyzes campaign dimensional patterns across hundreds of contexts builds intuition that compounds.

They start recognizing subtle signals—"this audience psychology pattern reminds me of something we saw 18 months ago in a completely different industry"—that pure pattern-matching algorithms miss.

They develop contextual judgment about when patterns transfer and when they don't. They know that dimensional similarity across buyer sophistication + competitive intensity + proof requirements might predict that challenger positioning works—but only if organizational complexity isn't too high and sales cycles aren't too long. That qualification comes from seeing the pattern succeed in some contexts and fail in others.

They build multi-disciplinary fluency. A strategist who's run enterprise software campaigns, financial services campaigns, healthcare campaigns, and manufacturing campaigns can recognize structural similarities that AI analyzing each domain independently wouldn't surface. They understand that what looks like "healthcare compliance" challenges often rhyme with "manufacturing quality assurance" challenges when you map them dimensionally—because both involve risk-averse buyers, regulatory scrutiny, and career-stakes decisions.

They accumulate edge case awareness. AI analyzing historical patterns identifies what worked on average. Humans who've lived through specific campaign failures remember the edge cases where standard approaches broke: "That dimensional pattern suggests aggressive competitive positioning would work, but we tried that in Q3 2023 when a regulatory change made buyers more conservative, and it backfired completely."

This institutional memory doesn't reset when AI models update.

When GPT-6 launches with better capabilities than GPT-5, the human expertise that guided GPT-5 orchestration transfers immediately to GPT-6. The pattern recognition about which campaigns rhyme, which insights transfer, which strategic approaches work—all of that persists.

But AI-dependent approaches reset partially with each model generation. Prompts that worked perfectly with GPT-5 might need complete restructuring for GPT-6.

Workflows optimized for one model's strengths might not leverage the next model's advantages. Strategic insights extracted from historical analysis might need re-validation.

Technical Systems That Persist Intelligence

Sophisticated organizations augment human memory with technical architecture designed to capture and retrieve pattern intelligence:

- Pattern libraries that document which dimensional combinations predicted success across specific contexts

- Structured knowledge bases that capture why certain strategic approaches worked or failed

- Retrieval-augmented generation (RAG) systems that surface relevant historical patterns when analyzing new campaigns

- Vector databases that enable semantic search across accumulated campaign intelligence

But—and this is critical—these technical systems are only as valuable as the human expertise that decides what patterns matter enough to capture, how to structure intelligence for retrieval, and when to trust historical patterns versus when context has shifted enough to override them.

The technical architecture enables persistent memory. Human judgment determines what's worth remembering and how to apply it.

Why This Creates a Sustainable Advantage

The combination of human institutional memory and technical persistence creates intelligence that compounds over time:

Year 1: You analyze 50 campaigns, extract dimensional patterns, and develop an initial intuition about what transfers.

Year 2: You analyze 100 more campaigns, refine pattern understanding, encounter edge cases that modify your frameworks, and strengthen your judgment about which patterns matter most.

Year 3: You analyze 150 more campaigns. Your pattern library now reflects 300 campaigns across diverse contexts, your team's intuition is genuinely sophisticated, and you're recognizing non-obvious rhymes that year-1-you would have missed.

This accumulation doesn't happen with AI alone. Generic AI tools don't build pattern libraries from your specific campaign experiences. Platform AI doesn't transfer learning across advertisers. Even sophisticated internal AI implementations reset partially when models update or team members turn over.

But human expertise augmented by persistent technical architecture compounds continuously. Each campaign analyzed strengthens pattern recognition. Each edge case encountered deepens contextual judgment. Each model advancement gets leveraged by expertise that persists across technology generations.

This expertise is powerful: it isn't just "knowing what worked before." It's developing judgment about why things worked, when patterns transfer, how context modifies applicability, and which dimensional combinations predict success in new situations.

That judgment—that sophisticated intuition built through years of diverse campaign exposure—is what allows you to look at a new campaign and say, "this rhymes with three campaigns we analyzed 18 months apart in completely different industries, and here's the strategic insight that transfers even though the surface details look nothing alike."

AI can surface mathematical similarity. Human institutional memory recognizes strategic rhyming.

That's the difference between pattern detection and pattern recognition.

And that's why institutional memory compounds competitive advantage while AI capabilities reset periodically.

Why Quality Control Becomes Harder as AI Gets Better

A counterintuitive dynamic emerges from the model improvements: as AI-generated outputs improve in quality, distinguishing good from great becomes harder, not easier.

When AI produces obviously mediocre content, quality control is straightforward. Bad writing stands out. Strategic errors are visible. You can quickly identify what needs fixing.

But as AI advances, outputs become increasingly polished. GPT-5 produces professional-quality copy. Claude Sonnet 4.5 generates sophisticated strategic reasoning. Gemini 3 creates well-structured analytical frameworks.

There's a problem with this: polished output that sounds sophisticated can still be strategically wrong.

AI can generate email copy that reads beautifully, follows brand voice guidelines perfectly, and includes all the right keywords—while completely missing what would actually resonate with risk-averse enterprise buyers making career-stakes decisions.

AI can produce competitive positioning that's logically structured, comprehensively argued, and backed by data—while fundamentally misunderstanding the psychological dynamics of how buyers evaluate challenger brands versus incumbents.

AI can create dimensional analysis that maps campaigns across all five dimensions with mathematical precision—while missing the contextual nuances that determine whether identified patterns will actually transfer to your specific situation.

This is where experienced marketing judgment becomes even more valuable as AI advances.

A junior marketer might read AI-generated strategic recommendations and think, "This sounds smart, let's do it." Inexperienced professionals lack the pattern recognition built through years of seeing what works and what doesn't to spot the subtle strategic errors hiding in sophisticated-sounding analysis.

An experienced strategist with 15+ years across diverse campaign contexts might read the same AI output and immediately recognize something: "This analysis correctly identifies dimensional similarity in terms of buyer behavior and competitive context, but it's missing that organizational complexity at this company size creates decision-making dynamics that will break this approach. I've seen this exact pattern fail twice before in similar contexts."

That recognition—that judgment about what's missing, what's contextually wrong despite sounding right, what will work in theory but fail in practice—is built through extensive exposure to diverse situations where you saw recommendations that sounded good get tested and fail.

It's pattern recognition about what doesn't work, which is often more valuable than pattern recognition about what does.

And it becomes more valuable, not less, as AI outputs become more polished and professional-looking.

The Insight Gap Between Sophisticated Outputs and Strategic Accuracy Will Widen as AI Improves

When AI was obviously limited, everyone knew to be skeptical. Now that AI produces genuinely impressive work, distinguishing "impressive" from "actually correct" requires deeper expertise.

This creates a dangerous dynamic for organizations that assume "better AI = less need for human oversight." The opposite is true. Better AI requires more sophisticated oversight because the errors are more subtle, and the confidence in outputs needs to be higher when they look more professional.

The Quality Control Moat Will Need to be Strengthened as AI Advances

Practitioners with deep experience across hundreds of campaigns in diverse contexts develop calibrated skepticism. They know where AI typically makes mistakes. They recognize the edge cases. They can spot strategic errors that junior practitioners miss because the writing quality is high.

This judgment isn't something you learn from a training course. It's built through years of seeing AI suggestions tested in the market—watching some succeed and others fail, and developing intuitions about which patterns actually transfer and which sound good but don't work.

And that intuition compounds over time, while AI capabilities reset with each model generation.

The Sustainable Moat

Now we can see why sophisticated pattern recognition creates a defensible competitive advantage that strengthens over time rather than eroding as AI advances:

The technology will continue advancing and commoditizing.

Whatever models exist today will be surpassed by better models next year. Whatever AI capabilities seem impressive now will become table stakes. The technology itself is not the moat.

Orchestration expertise compounds.

Knowing how to use AI effectively for sophisticated strategic work will continue to be crucial—which models to use for which tasks, how to structure prompts for reliable outputs, when to trust AI suggestions versus when to override with human judgment, and how to validate pattern recognition computationally while applying contextual expertise. This expertise doesn't commoditize because it requires ongoing judgment about a constantly changing landscape.

Pattern intelligence accumulates.

Every campaign analyzed adds to dimensional understanding. Every edge case encountered strengthens contextual judgment. Every cross-domain pattern recognized enriches the intelligence that informs future campaigns. This creates a flywheel, where more campaigns lead to better insights, leading to better campaigns, leading to more campaigns.

Multi-disciplinary expertise is rare.

Most marketing professionals specialize in specific domains—SaaS marketing, healthcare marketing, enterprise ABM, demand generation. Practitioners with deep expertise across multiple domains, multiple channels, and multiple buyer contexts are genuinely scarce. And that cross-domain expertise is exactly what's required for sophisticated dimensional pattern recognition.

Institutional memory persists and grows.

While AI models reset with each conversation and partially reset with each model generation, experienced humans carry forward pattern recognition built across years of diverse campaign contexts, augmented by technical systems designed to persist and retrieve pattern intelligence. This accumulated judgment about what works, what transfers, what matters—this is the sustainable moat.

Put these together, and you get a competitive advantage that doesn't erode as AI advances. *It appreciates.*

Generic AI tools will continue to commoditize every 6–12 months. They'll get better and cheaper. But they won't develop pattern intelligence, because they don't analyze actual campaign performance across diverse contexts. They won't develop institutional memory, because they reset with each user interaction.

Platform AI will continue optimizing for platform revenue rather than advertiser outcomes. It will get more sophisticated at maximizing your spend while maintaining just enough performance to prevent churn. But it will never develop objective strategic intelligence, because its incentives are fundamentally misaligned.

Black box AI systems will continue failing at strategic work that requires contextual judgment, explainability, and adaptation as market conditions shift. They'll get better at optimization within narrow constraints but won't develop the sophisticated judgment required for dimensional pattern recognition.

Sophisticated practitioners orchestrating AI with dimensional pattern intelligence will continue pulling ahead. Each campaign they analyze enriches their pattern recognition. Each AI model advancement they integrate strengthens their capabilities. Each edge case they encounter deepens their judgment. Each architectural improvement enhances their ability to leverage technology advances.

The moat isn't the technology. *The moat is the expertise orchestrating the technology.*

And that expertise becomes exponentially more valuable as AI capabilities expand—because the leverage each expert decision creates grows proportionally with AI's execution capabilities.

This is the inversion of the *Terminator* premise:

AI advancement doesn't make experienced marketers obsolete.

It makes marketers who can orchestrate AI sophisticatedly far more valuable.

It makes marketers who understand dimensional patterns increasingly essential.

It makes marketers with multi-disciplinary expertise across diverse contexts increasingly rare and valuable.

The gap between sophisticated AI orchestration and naive AI use is widening rapidly. And it's defined primarily by depth of marketing expertise, breadth of cross-domain experience, and judgment about what actually works—not by access to technology.

What This Means for Pattern Recognition

Let's return to the core argument of this book:

Multi-dimensional pattern recognition that finds campaigns that genuinely rhyme requires exactly this combination of sophisticated AI capabilities and deep human expertise.

The AI enables the computational complexity—analyzing hundreds of campaigns across dozens of dimensions, calculating similarity scores, surfacing non-obvious rhymes that human cognition alone couldn't systematically process.

The human expertise provides the strategic judgment—determining which dimensions matter for specific questions, validating that calculated similarities represent genuine strategic opportunities rather than spurious correlations, interpreting what dimensional patterns mean for campaign strategy, and recognizing when to trust patterns versus when context overrides historical data.

The technical architecture provides persistent memory and retrieval systems that compound intelligence over time rather than resetting with each model generation or team change.

This is why dimensional pattern recognition isn't easily replicated.

You can't just buy access to AI models and expect sophisticated pattern intelligence. The models are commodities. Everyone has access.

You can't just hire smart people and expect them to figure it out quickly. The expertise required comes from years of exposure to diverse campaign contexts, learning which patterns transfer and which don't, and developing calibrated judgment about what works.

You can't just build it once and expect it to stay competitive. The AI landscape evolves constantly. Model capabilities change. New approaches emerge. Maintaining sophistication requires an ongoing investment in understanding and integrating advances.

But if you combine multi-disciplinary marketing expertise with sophisticated AI capabilities, persistent technical architecture for pattern intelligence, and continuous learning across diverse campaigns, you create something genuinely defensible.

This defensibility doesn't come about because competitors can't access the same technology. They can.

It's not defensible because the approach is secret or proprietary. The conceptual framework is right here in this book.

It's defensible because executing this approach well requires exactly the combination of depth, breadth, judgment, and institutional memory that takes years to build and continues compounding over time.

That's the human-AI moat.

And it's why, counter to the *Terminator* narrative, AI advancement makes experienced marketing expertise more valuable, rather than less valuable.

The better AI gets at execution, the more each strategic decision matters.

The more options AI generates, the more valuable judgment becomes about which options actually work.

The more sophisticated AI outputs look, the harder quality control becomes—and the more valuable experienced pattern recognition is for catching subtle strategic errors.

The more campaigns you analyze with this approach, the richer your pattern intelligence becomes.

This advantage appreciates over time, rather than depreciating.

Which means building it now—whether by developing internal capability or partnering with those who've already built it—creates leverage that strengthens with each campaign, each model advancement, and each year of accumulated institutional memory.

FINDING RHYMES—THE HUMAN-AI MOAT

> **Author's Note:** The human-AI moat deserves a deeper exploration than this chapter allows. The dynamics examined here—why expertise appreciates as AI advances, how institutional memory compounds, where sustainable advantages form in AI-saturated markets—have implications far beyond marketing campaign intelligence. If there's sufficient interest, a future work will examine this phenomenon in detail across industries and use cases. For now, we'll focus on why this moat specifically enables dimensional pattern recognition at scale.

In the next chapter, we'll explore the technical architecture that makes dimensional pattern recognition practical—how embedding models work, how similarity algorithms function, how persistent intelligence systems operate, and why these technical capabilities enable insights that weren't accessible even a few years ago.

But remember: the technical architecture is only valuable because of the human expertise guiding it.

The AI provides computational muscle. The expertise provides strategic judgment. The architecture provides persistent memory.

And in sophisticated strategic work, judgment informed by compounding memory is what creates a sustainable competitive advantage.

Chapter Eight

How AI Enables Pattern Recognition at Scale

THE BREAKTHROUGH ISN'T THAT AI can think. It's that AI can represent complex strategic concepts mathematically—and that changes everything about finding patterns.

When you describe a campaign strategy in natural language—"we targeted skeptical mid-market buyers who'd been burned before with trust-building content that acknowledged their concerns before presenting solutions"—that's rich strategic information. But it's unstructured. There's no inherent way to calculate how similar that campaign is to thousands of others, or to surface which historical campaigns share the most relevant patterns.

Modern AI solves this through a capability that sounds technical but has profound strategic implications: it can convert qualitative campaign descriptions into mathematical representations that preserve strategic meaning and relationships.

This makes dimensional pattern recognition computationally tractable at scale.

Let me show you how this actually works, and why it matters.

From Strategy to Vectors: The Embedding Breakthrough

Traditional databases store campaigns as structured fields: Industry. Channel. Audience. Performance metrics. You can filter and query these fields, but comparing campaigns requires matching on explicit attributes. "Show me B2B SaaS email campaigns" works fine. "Show me campaigns that share similar strategic DNA across multiple subtle dimensions" doesn't.

Embedding models change the game entirely.

An embedding transforms complex information—campaign strategy descriptions, audience psychology notes, competitive context analysis—into a point in high-dimensional mathematical space. Not as a simplification, but as a rich numerical representation that captures nuanced meaning.

Think of it like this: every strategic concept, every audience insight, every positioning choice gets mapped to coordinates in a space with hundreds or thousands of dimensions. Campaigns that share similar strategic characteristics cluster together in this space. Campaigns that differ significantly sit far apart.

This isn't categorization. This is continuous dimensional mapping.

The power becomes clear when you realize what this enables: you can mathematically calculate which campaigns occupy similar strategic territory across any combination of dimensions that matter for your specific question.

Looking for campaigns that rhyme on audience skepticism + competitive intensity + proof strategy? The mathematical representation lets you find the nearest neighbors across exactly those dimensions.

Need campaigns that match on buyer sophistication + resource constraints + message complexity? Different dimensional slice, same mathematical approach.

The embedding model doesn't just store campaign attributes separately—it creates a unified representation where dimensional relationships are preserved and queryable.

This is what makes "campaigns that rhyme" practically discoverable rather than just theoretically interesting.

Similarity at Scale: Beyond Simple Matching

Once campaigns exist as mathematical representations, measuring similarity becomes systematic rather than intuitive.

There are multiple mathematical approaches to calculating how "close" two campaigns are in high-dimensional space. Some measure geometric distance between points. Others calculate the angle between vectors. Still others evaluate how campaigns cluster in relation to specific dimensions.

The specific mathematics matter less than what they collectively enable: the ability to ask sophisticated strategic questions and get computationally derived answers.

"Which campaigns in our pattern library share the strongest dimensional similarity with this new initiative we're planning?"

"Of the campaigns that rhyme on audience psychology and competitive context, which ones also succeeded with similar resource constraints?"

"Show me campaigns that match on three specific dimensions while differing on others—I want to isolate what changes when dimension X varies."

This is strategic intelligence that simply wasn't accessible before. Not because marketers couldn't recognize patterns intuitively, but because the computational infrastructure to systematically surface and validate patterns at scale didn't exist.

The similarity calculations provide candidates—campaigns worth examining more closely because they share strategic DNA. But here's what's critical: *the calculations don't interpret what the patterns mean.*

They don't determine which insights transfer. They don't explain why certain dimensional combinations predict success. They don't know how to apply the patterns strategically.

They surface mathematical similarity. Human expertise evaluates strategic relevance.

The experienced judgment we explored in Chapter 7—knowing which patterns actually matter, understanding contextual factors that modify how patterns transfer, recognizing when calculated similarity represents genuine strategic rhyming versus spurious correlation—that's what transforms computational pattern detection into actionable intelligence.

Orchestrating Multiple AI Capabilities

Pattern recognition at scale requires more than just finding similar campaigns. It requires multiple distinct capabilities working together: strategic reasoning about which dimensions matter, creative generation of campaign variations, analytical synthesis of performance data, logical validation of pattern interpretations, and systematic questioning of assumptions.

No single AI model excels at everything.

Different models have different architectural strengths. Some reason more effectively about strategic context. Others generate more creative variations. Still others synthesize structured analysis better or maintain longer contextual memory. (We touched on this briefly in "A Brief Technical Preview" in Chapter 4, and it was also addressed in "The Multi-LLM Orchestration Advantage" in Chapter 7.)

Multi-model orchestration means using different AI capabilities for what they do best, then combining those capabilities systematically.

This isn't about having "backup AI" in case one fails. It's about recognizing that dimensional pattern recognition has genuinely different computational requirements at different stages, and different models handle those requirements with different levels of sophistication.

Strategic analysis of which dimensions matter for a specific question requires different capabilities than generating 50 creative variations of a message strategy. Performance data synthesis requires different capabilities than validating the logical consistency of pattern interpretations.

When you orchestrate multiple models—routing different analytical tasks to models with relevant architectural strengths, cross-validating insights across models, using ensemble approaches where models vote on pattern significance—you get more robust intelligence than any single model provides.

There's an additional advantage that's easy to miss: resilience to model evolution.

AI capabilities change rapidly. Models get updated, deprecated, and replaced. New models emerge with different strength profiles. If your entire pattern intelligence infrastructure depends on a single model, you're vulnerable to regression when that model changes or loses capabilities in updates.

Multi-model orchestration creates architectural flexibility. When one model's capabilities shift, others compensate. When new models emerge with novel strengths, they can be integrated without rebuilding the entire system.

This matters more than it might seem. The AI landscape evolves on 6–12-month cycles. What's cutting-edge today becomes outdated quickly. Systems built around single models become brittle. Systems built around orchestrated capabilities adapt.

Orchestration requires judgment about which capabilities to use when, in what sequence, with what validation.

That's not something AI does inherently. This kind of orchestration requires expertise about how different analytical tasks require different computational approaches, which models handle which challenges effectively, how to validate cross-model insights, and when to trust ensemble agreement versus when disagreement signals the need for human interpretation.

Persistent Intelligence: How Pattern Knowledge Compounds

There's a final technical capability that makes dimensional pattern recognition sustainable rather than ephemeral: systems that persist and retrieve pattern intelligence over time.

Standard AI interactions reset with each conversation. The model doesn't remember previous analyses, accumulate insights from past campaigns, or build on pattern recognition from earlier work. Each query starts fresh.

For pattern intelligence, this is a critical limitation. Valuable insights emerge from analyzing hundreds or thousands of campaigns, recognizing which dimensional combinations predict success, and understanding which patterns transfer across contexts and which don't. That knowledge should compound—each new campaign analyzed should enrich the pattern library, making future pattern recognition more sophisticated.

This is where retrieval-augmented generation architectures, vector databases, and knowledge graphs become strategically important.

These aren't generic AI tools. They're systems designed specifically to persist dimensional campaign representations, maintain relationships between campaigns that share strategic patterns, and enable sophisticated retrieval based on multi-dimensional similarity.

When you analyze a new campaign and want to find which historical campaigns rhyme most strongly, you're not recalculating similarity against the entire campaign corpus from scratch each time. You're querying a persistent intelligence system that's been purpose-built to surface relevant patterns based on dimensional proximity.

The system remembers which campaigns exist in the pattern library. It maintains their dimensional representations. It enables fast retrieval of campaigns that match specific dimensional profiles. It supports queries like "find campaigns similar on dimensions 1, 3, and 5 but different on dimension 2—I want to isolate what changes when competitive intensity shifts."

This persistent architecture is what allows pattern intelligence to accumulate rather than reset.

Each campaign added to the library enriches future pattern recognition. Each validation of which patterns transfer compounds the system's sophistication. Each refinement of which dimensional combinations predict success makes subsequent analyses more valuable.

But—and this is absolutely critical—the value of persistent intelligence depends entirely on what you persist and how you organize it.

A vector database storing campaign embeddings is just technology. The strategic value comes from knowing which campaign characteristics to capture, how to structure dimensional relationships, which metadata preserves contextual nuance, and how to weight similarity across different dimensional combinations.

That's encoded marketing expertise, not generic technical infrastructure.

It's why building sophisticated pattern intelligence systems requires both deep marketing domain knowledge and technical architectural sophistication. The technology enables persistence and retrieval at scale. The expertise determines what's worth persisting and how to structure it for strategic usefulness.

Transparency and Auditability: Not a Black Box

One concern people raise about AI-powered marketing intelligence is opacity: "How do I know the AI isn't just making things up? How can I trust recommendations I can't validate?"

This is a legitimate concern with many AI applications. But dimensional pattern recognition, done correctly, is fundamentally transparent and auditable.

When the system suggests that Campaign A rhymes with Campaign B, you can ask: "Show me the dimensional analysis. Which dimensions are similar? What are the similarity scores? What specific characteristics create the pattern match?"

The system can show you things like the following:

- Campaign B's dimensional profile across all five core dimensions

- The similarity scores for each dimension

- Which specific sub-dimensions drive the overall match

- How Campaign B performed in its context

- What strategic choices Campaign B made that might inform your approach

You're not getting a black box recommendation: "Our AI says do this." You're getting transparent pattern matching: "Here's a campaign that shares your dimensional profile, here's specifically what's similar, here's how it performed, and here's what might transfer."

You can interrogate every suggested rhyme. Question why the system thinks campaigns are similar. Override suggestions when your judgment says the mathematical similarity doesn't represent strategic relevance. Refine the dimensional weighting for future analyses based on what you learn.

This transparency is essential for sophisticated strategic work.

You can't make high-stakes campaign decisions based on opaque AI recommendations. You need to understand the reasoning, validate the logic, and apply contextual judgment about whether the pattern actually applies to your situation.

Dimensional pattern recognition provides that transparency. The calculations are explainable. The dimensional analysis is visible. The pattern matching is auditable.

This doesn't mean the system is always right. It means that when it's wrong, you can figure out why—maybe it weighted a dimension incorrectly for your context, maybe it missed a contextual factor that breaks the pattern, maybe the similar campaign succeeded for reasons that won't transfer.

That feedback loop—where human expertise validates, refines, and improves the system's pattern recognition over time—is how intelligence compounds.

The Orchestra Conductor Metaphor

Let's pull this together with the metaphor that best captures how modern AI enables pattern recognition at scale.

Think of a symphony orchestra. You have strings, brass, woodwinds, percussion—each section with unique capabilities and sounds. No single section can perform the entire composition. The power comes from orchestration: knowing which instruments play when, how they harmonize, and when certain sections lead while others support.

The conductor doesn't play any instrument. But the conductor's expertise—understanding the composition, knowing each section's strengths, coordinating timing and dynamics—is what transforms individual instrumental capabilities into coherent musical expression.

Dimensional pattern recognition works the same way.

You have multiple AI capabilities: embedding models that create mathematical representations, similarity algorithms that find patterns, language models that reason about strategy, systems that generate creative variations, and architectures that persist and retrieve intelligence.

No single capability delivers complete pattern recognition. The power comes from orchestration: routing strategic reasoning to models strong at contextual analysis, using different models for creative generation versus analytical synthesis, cross-validating pattern significance across multiple approaches, and persisting insights in structured systems that enable sophisticated retrieval.

And just like the conductor doesn't play instruments, but is essential for coherent performance, experienced marketing expertise doesn't execute the computational work, but is essential for strategic orchestration.

The expertise of the conductor determines a number of key things:

- Which dimensions matter for the specific strategic question

- How to weight dimensional similarity for different contexts

- When calculated patterns represent genuine strategic rhymes versus spurious correlations

- Which AI capabilities to use for different analytical tasks

- How to validate cross-model insights

- What to persist in the intelligence system and how to structure it

- How to translate dimensional patterns into an actionable campaign strategy

The AI provides the instrumental capabilities. The expertise provides the orchestration that makes those capabilities produce coherent strategic intelligence.

Neither works alone. Together, they make dimensional pattern recognition at scale practically possible—turning what was theoretical a few years ago into an operational reality today.

We've now covered how dimensional pattern recognition works conceptually (Chapters 4–6), why it requires human expertise plus AI to create a defensible moat (Chapter 7), and how the technology actually enables it at scale (this chapter).

The next question is this: how do you protect your competitive advantage and client confidentiality when building pattern intelligence across multiple campaigns?

That's what we'll explore in Chapter 9.

Chapter Nine

Privacy-First Intelligence

WE'VE ESTABLISHED WHAT PATTERN recognition is, why it requires human expertise orchestrating AI, and how the technology enables dimensional analysis at scale.

But there's a critical question we haven't fully addressed—one that could undermine confidence in the entire approach:

How do you build pattern intelligence within an agency across multiple campaigns without compromising competitive advantage or client confidentiality?

This isn't a minor operational concern. It's fundamental. If pattern recognition requires accessing proprietary campaign details, creative strategies, or performance specifics that reveal competitive intelligence, the approach becomes a non-starter for most marketers. You can't build better intelligence by compromising the very advantage that intelligence is supposed to create.

The answer lies in understanding what pattern recognition actually needs versus what it specifically doesn't need—and why that distinction isn't just about protecting privacy, but about creating better intelligence.

What Pattern Recognition Actually Requires

Most people assume that dimensional pattern analysis needs detailed campaign information to work. Company names. Product specifics. Creative assets. Exact targeting parameters. Detailed performance metrics tied to identifiable campaigns. But this is a misunderstanding: it doesn't.

What pattern recognition requires are the dimensional characteristics and aggregate performance outcomes. Not the campaign specifics.

The dimensional signature *is* the intelligence.

In this chapter, we'll go deeper than simple campaign comparison. Pattern recognition captures something far more sophisticated: message sequences, channel orchestration, timing cadence, and progressive narrative arcs across multiple touchpoints.

Consider two campaigns:

Campaign A (healthcare compliance SaaS):

- Week 1: Email (trust-building message: "Healthcare data breaches increased 47% in regulated environments")

- Weeks 1–2: Display retargeting (social proof: "Join 200+ hospitals protecting patient data," three impressions, strategic frequency capping)

- Week 2: LinkedIn Sponsored Messaging (outcome-focused: "Reduce compliance risk in 90 days with automated monitoring")

- Week 3: Video pre-roll (urgency: "New HIPAA requirements take effect Q1," two impressions)

- Result: 240 leads, 68% director-level or above, 72% near-term intent, conversion spike in Week 3

Campaign B (financial services risk management):

- Week 1: Email (trust-building message: "Regulatory penalties reached $2.8B across financial services")

- Weeks 1–2: Display retargeting (social proof: "Trusted by 150+ financial institutions," three impressions, matched frequency capping)

- Week 2: LinkedIn Sponsored Messaging (outcome-focused: "Achieve audit readiness in 60 days with real-time tracking")

- Week 3: Video pre-roll (urgency: "Securities and Exchange Commission (SEC) reporting deadline approaching," two impressions)
- Result: 180 leads, 71% VP-level or above, 73% near-term intent, conversion spike in Week 3

Campaign specifics that differ:
- Industries (healthcare vs. financial services)
- Products (HIPAA compliance vs. risk management)
- Creative executions (specific headlines, imagery, copy)
- Exact targeting parameters
- Budget allocations
- Specific companies generating leads

Dimensional pattern both share:
- Risk-averse executives in regulated industries
- Competitive mature markets
- Trust-building → social proof → outcome → urgency message progression
- Email → display → LinkedIn → video channel sequence
- Three-week conversion cadence
- Progressive message intensification
- Cross-channel narrative reinforcement
- High percentage of senior-level leads with strong intent signals

The pattern that transfers across these campaigns has nothing to do with HIPAA penalties or SEC deadlines. It has everything to do with the dimensional characteristics:

who you're targeting (risk-averse buyers in regulated contexts), how you're messaging them (trust-building foundation building to urgency), what sequence you're using (progressive intensification, not random message rotation), what timing matters (three-week cadence with strategic gaps, not constant bombardment), and how channels reinforce each other (each touchpoint building narrative momentum).

A third campaign targeting risk-averse procurement executives at manufacturing companies could benefit from this pattern without knowing anything about Campaign A or Campaign B's specific details. The dimensional intelligence transfers—message progression, channel sequencing, timing cadence, narrative arc. The campaign specifics don't—creative execution, targeting parameters, budget allocation, specific results.

This isn't just about privacy protection. It's about what actually creates transferable intelligence. The message sequence pattern "trust → proof → outcome → urgency over three weeks" is valuable across contexts. The specific headline, "Healthcare data breaches increased 47%," is not.

Why Campaign Specifics Are Actually Noise

There's a counterintuitive insight at the heart of privacy-first pattern recognition:

Forcing abstraction to the dimensional level creates better intelligence than storing detailed campaign specifics.

When you have access to complete campaign details, you risk overfitting to irrelevant specifics that don't transfer across contexts. You might conclude "HIPAA penalty messaging drives high engagement" when the actual transferable pattern is "consequence-focused messaging works for risk-averse audiences in regulated industries." The first insight is narrow and non-transferable. The second is dimensional and broadly applicable.

By extracting only dimensional characteristics and aggregate outcomes, you're forced to identify the patterns that actually matter—the structural similarities that transcend industry verticals, product categories, and creative executions.

Think about medical research. Individual patient details are rigorously protected. Researchers never know "Patient #427 responded well to Treatment B." But they do learn "patients with these physiological characteristics respond better to Treatment B than Treatment A." The abstraction to the pattern level—forced by privacy requirements—is precisely what makes the research transferable to new patients.

The same principle applies to campaign intelligence. You don't need to know that "Client X's CISO campaign using [specific creative] achieved 28% engagement." You need to know that "campaigns targeting risk-averse technical decision-makers in competitive markets achieve 30% higher engagement with trust-building messaging approaches versus feature-focused approaches."

Remember from Chapter 2: with enough data, you can torture numbers into confessing to anything. You could probably find a pattern showing that "senior executives with last names starting with G respond better to blue calls to action on Tuesdays." That's not insight—that's overfitting to statistical noise.

Privacy constraints force you away from this trap. When you can't store campaign specifics, you can't overfit to irrelevant correlations. You're forced to abstract to dimensional patterns that actually transfer across contexts.

The dimensional abstraction is more valuable than the campaign-specific details because it's transferable to new contexts. And privacy requirements force that valuable abstraction.

Let me say this again, because it's counterintuitive enough that it bears repeating: *privacy requirements make you a better marketer.* They force you to think dimensionally instead of copying surface-level tactics. They push you toward transferable patterns instead of non-transferable specifics. They make you abstract to what actually matters instead of overfitting to what doesn't.

Most marketers treat privacy like a constraint to work around. It's actually a feature that improves your thinking.

< insert head exploding emoji (or wide-eyed emoji if preferred) here >

Architecture That Can't Compromise What It Doesn't Capture

Privacy-first pattern recognition isn't about restricting access to sensitive data that exists in a database. It's about architecting systems that extract dimensional patterns without capturing campaign specifics in the first place.

The distinction matters enormously.

Privacy through access control says: "We have your detailed campaign data, but we promise to restrict who can see it." That's not privacy by design. That's privacy by policy. Policies can change. Access controls can fail. Human error happens.

Architectural privacy says: "The system is designed to extract dimensional patterns and aggregate outcomes. It can't expose campaign specifics because it never captured them." You can't leak what you never stored. You can't reverse-engineer from abstractions that never included the specifics.

When a campaign enters pattern analysis, what gets extracted is the dimensional signature and performance outcomes:

- Audience attributes: risk-averse executives, regulated industry context, technical decision-makers

- Market context: competitive mature market, multiple established alternatives, complex procurement

- Message strategy: trust-building foundation, progressive intensification (trust → proof → outcome → urgency), consequence-oriented framing

- Channel orchestration: email → display → LinkedIn → video sequence, three-week conversion cadence, strategic frequency capping, narrative arc across touchpoints

- Execution quality: sophisticated targeting, strong creative consistency, professional production, coordinated timing

- Aggregate lead characteristics: 60% director-level or above, 72% near-term intent signals

- Performance outcomes: 8.3% conversion rate, 240 qualified leads, 34% engagement lift versus baseline, Week 3 conversion spike

What doesn't get captured:
- Company name or identifiable client information
- Product specifics or proprietary positioning
- Actual creative assets, headlines, or messaging copy
- Specific targeting parameters that reveal competitive intelligence
- Individual prospect names, companies, or contact information
- Budget allocations or media spend details
- Any details that would allow reverse-engineering of the specific campaign

The dimensional coordinates and aggregate outcomes are what enable pattern recognition. Everything else is not just unnecessary—it's actively counterproductive because it creates overfitting risk and privacy exposure without adding intelligence value.

The Two-Layer Privacy Framework

Sophisticated pattern intelligence actually requires two distinct privacy layers, each serving different purposes:

Campaign-level anonymization

Client identities, product specifics, creative details, and proprietary strategies are abstracted to dimensional patterns. This protects the competitive advantage. Campaign A and Campaign B can't reverse-engineer each other's approaches from knowing they share dimensional signatures. The pattern that transfers is "risk-averse buyers in regulated contexts respond better to trust-building message progressions with three-week cadence"—not the specific creative execution, product positioning, targeting parameters, or budget allocation that either campaign used.

Lead-level aggregation

Individual prospect information never enters the pattern intelligence system. When campaigns generate leads through content syndication, gated landing pages, or lead generation advertising, those leads contain personal information—names, titles, companies, email addresses, intent signals from custom questions. That information is handled in operational systems governed by appropriate privacy regulations and client agreements. What enters pattern analysis is aggregated: "Campaign generated leads with 60% director-level titles, 72% near-term buying intent, primarily from mid-market financial services companies." The pattern system learns about lead quality characteristics without accessing individual prospect data.

This architectural separation means pattern intelligence operates on two levels of abstraction simultaneously: campaigns abstracted to dimensions, and leads abstracted to aggregate characteristics.

Neither layer requires—nor benefits from—storing the sensitive details that create privacy risk.

Why This Isn't Just Platform Benchmarks with Better Marketing

At this point, you might be thinking: "Wait, don't advertising platforms already do this? LinkedIn publishes 'Trends Influencing Higher Education Marketing.' Meta releases 'B2B Marketing Benchmarks.' Google shares industry performance data. Isn't this just the same thing?"

No. And understanding the distinction is critical.

Platform benchmark reports aggregate at the surface category level. They say things like the following:

- "Higher education display ads averaged 0.8% CTR in Q3"

- "B2B SaaS email campaigns achieved 22% open rates"

- "Financial services video completion rates: 45%"

These are exactly the meaningless industry averages we've been criticizing throughout this book. They group together incomparable campaigns based on crude categorization—industry vertical, channel type, maybe company size—without any dimensional nuance.

Dimensional pattern recognition operates at a fundamentally different level of sophistication.

Instead of "Higher Ed averaged 0.8% CTR," you get: "Campaigns targeting risk-averse academic administrators in competitive funding environments achieved a 2.1% CTR with outcome-focused messaging versus 0.7% with feature-focused messaging, with progressive message sequencing (trust → proof → outcome → urgency) outperforming random message rotation by 3x."

The platform reports tell you what happened across a crude category. Dimensional pattern recognition tells you what dimensional characteristics correlate with performance outcomes—and why similar patterns might transfer to your context.

Platform benchmarks are descriptive statistics about past performance in surface categories. "Here's what happened on average to campaigns sort of like yours."

Dimensional pattern recognition is analytical intelligence about transferable patterns. "Here's what dimensional characteristics correlate with performance outcomes, abstracted to a level that helps you make better strategic decisions for your specific context."

Platform benchmark: "B2B SaaS companies got 22% open rates."

Dimensional pattern: "Campaigns targeting risk-averse technical buyers in competitive markets achieved 34% open rates with trust-building subject lines versus 18% with feature-focused subject lines, with this performance advantage holding across SaaS, healthcare IT, and financial services technology verticals."

See the difference? One tells you an average that might not apply to you. The other tells you a dimensional pattern with enough contextual nuance that you can evaluate whether it's relevant to your situation—and adapt it intelligently.

Platform benchmarks often require sharing your campaign performance data with the platform to get included in their aggregations. You're trading your data for access to everyone else's averaged data.

Dimensional pattern recognition extracts patterns without requiring data sharing agreements or participation fees. The architectural approach—extracting dimensional characteristics and aggregate outcomes rather than storing campaign specifics—means you're not contributing your data to a pool. You're having dimensional patterns extracted that enable better strategic intelligence without compromising your competitive advantage.

Platform benchmarks with AI are still just sophisticated averaging of incomparable campaigns. Dimensional pattern recognition is a fundamentally different analytical approach.

What This Means for Competitive Protection

Let's address the core concern directly: How does building pattern intelligence across campaigns protect rather than compromise your competitive advantage?

The answer comes back to what actually transfers versus what stays proprietary.

Here are some examples of the kind of things that transfer through pattern recognition:

- "Trust-building messaging outperforms feature-focused messaging for risk-averse audiences in competitive markets."

- "Progressive message sequencing (trust → proof → outcome → urgency) generates higher engagement than random message rotation."

- "Multi-touch sequences with an email → display → LinkedIn → video progression and a three-week cadence generate higher senior-level engagement than single-channel or constant-frequency approaches."

- "Campaigns targeting mid-market financial services with near-term intent signals convert at 3x higher rates with outcome-focused creative."

What remains proprietary:

- Your specific creative execution and messaging copy
- Your exact targeting parameters and audience definition
- Your product positioning and differentiation strategy
- Your budget allocation and media mix decisions
- Your competitive intelligence about market dynamics
- Your specific channel partnerships and vendor relationships
- The specific results you achieved with your specific campaign

If a competitor learns that "trust-building message progressions work better than random messaging for risk-averse buyers," they haven't learned your strategy. They've learned a dimensional pattern that they'd still need to execute with their own creative work, their own positioning, their own targeting, their own budget, their own orchestration, and their own timing.

Pattern intelligence tells you *what* tends to work in similar dimensional contexts. It doesn't tell you *how* any specific company executed their approach, what specific creative they used, or what specific results they achieved.

That's the protection built into dimensional abstraction. You can't reverse-engineer Campaign A from knowing it shares dimensional characteristics with Campaign B. You can only learn that campaigns with those dimensional characteristics tend to perform better with certain strategic approaches—which you then have to execute with your own proprietary implementation.

Your competitive advantage stays yours. What gets extracted are the patterns that inform better strategic decisions across similar contexts.

The Ethical Framework

There's a useful analogy to medical research that clarifies the ethical model here.

Medical researchers conduct studies across thousands of patients. Individual patient details are rigorously protected through privacy protocols—you can't identify which specific patients participated or what their individual outcomes were. But aggregate patterns advance treatment for everyone: "patients with these characteristics respond better to this treatment approach than to that alternative."

The ethical framework balances two imperatives: protect individual patient privacy absolutely, but enable collective learning that improves outcomes for future patients. You can't compromise the first to achieve the second. And you don't have to—proper anonymization and abstraction allow both.

Campaign pattern intelligence works the same way. Individual campaign details are protected through dimensional abstraction. But aggregate patterns improve strategic decisions across contexts: "campaigns with these dimensional characteristics achieve better outcomes with this strategic approach."

The ethical balance: protect competitive advantage and client confidentiality absolutely, but enable pattern learning that improves campaign performance across contexts.

And just like medical research, the architectural approach makes both possible simultaneously. You don't compromise protection to gain intelligence. Proper abstraction delivers both.

Why This Becomes More Important, Not Less

Here's the structural reality that makes privacy-first architecture essential rather than optional:

The web is getting more private, not less private.

Browser changes are eliminating third-party cookies. Privacy regulations are expanding globally. User expectations around data protection are increasing. Platform policies are restricting data access. Every structural trend points toward more privacy constraints, not fewer.

Marketing intelligence approaches that depend on accessing detailed user behavior, tracking individuals across properties, or aggregating personal information are swimming against an irreversible current. They might work today. They're obsolete tomorrow.

Pattern recognition based on dimensional characteristics and aggregate outcomes doesn't fight this trend—it's designed for it. The intelligence comes from campaign-level patterns that don't require individual tracking, personal data collection, or behavioral surveillance.

This isn't just about being virtuous. It's about being structurally aligned with where privacy requirements are headed. Systems built on accessing detailed personal or campaign data have an expiration date. Systems built on dimensional pattern abstraction are privacy-native by design.

What This Enables

Privacy-first architecture doesn't limit what pattern recognition can accomplish. It defines what makes pattern recognition valuable in the first place.

When you build intelligence on dimensional abstractions rather than campaign specifics, you create insights that transfer across contexts. When you aggregate lead characteristics rather than storing individual prospect data, you learn about quality patterns without privacy exposure. When you design systems that can't capture what they don't need, you eliminate entire categories of risk.

The result is intelligence that's both more protective and more valuable than approaches that require accessing sensitive details.

You learn that campaigns targeting risk-averse executives in competitive markets perform 30% better with trust-building message progressions—without compromising any client's specific execution or competitive strategy.

You learn that multi-touch orchestration with email → display → LinkedIn → video sequences and strategic timing cadence generates higher engagement from senior decision-makers—without exposing anyone's targeting parameters, budget allocation, or vendor relationships.

You learn that campaigns generating leads with strong near-term intent signals convert at 3x higher rates—without accessing individual prospect names, companies, or contact information.

You learn that progressive message sequencing outperforms random message rotation by 3x across risk-averse audiences—without revealing anyone's specific creative execution or messaging copy.

The patterns that matter are dimensional. The details that require protection are specific.

Privacy-first architecture aligns these perfectly: extract the dimensional patterns that create transferable intelligence; protect the specific details that create competitive advantage and privacy risk.

We've now covered how pattern recognition works conceptually, why it requires human expertise to create a defensible competitive advantage, how AI technology enables it at scale, and how privacy-first architecture makes it both ethical and protective.

The next question is this: if you're evaluating approaches to campaign intelligence—whether building internally or partnering externally—how do you assess whether an approach actually delivers sophisticated pattern recognition versus just repackaging generic benchmarks with AI buzzwords?

That's what we'll explore in Chapter 10.

Chapter Ten

Evaluating Pattern Recognition Approaches

THE PATTERN RECOGNITION OPPORTUNITY is becoming clear to marketers. The frustration with generic benchmarks is widespread. The promise of AI-native campaign intelligence is compelling.

Which means the market is about to get crowded with agencies and vendors claiming they can deliver it.

Some will be legitimately sophisticated. Most will be repackaging the same benchmark aggregation with new language. A few will be actively misleading—slapping "AI-powered pattern recognition" labels on tools that are neither intelligent nor recognizing actual patterns.

If you're looking for an agency (or a job at one), your challenge isn't finding one who claims to do pattern recognition. It's identifying the rare few who can actually deliver it.

This chapter gives you the framework to make that distinction. Not as a checklist that guarantees the right answer, but as a set of signals that separate sophisticated approaches from superficial ones. Think of this as developing pattern recognition for evaluating pattern recognition approaches.

Which is appropriately meta for a book about patterns.

The AI Washing Problem

AI capabilities are being wildly overstated across marketing technology right now.

Every marketing automation platform suddenly has "AI-powered optimization." Every content tool claims "AI-driven insights." Every analytics dashboard touts "machine learning recommendations." The AI label has become meaningless precisely because it's become ubiquitous.

This creates a specific problem for pattern recognition. The concept sounds sophisticated enough that vendors can claim it without understanding what it actually requires. They're not necessarily lying—they might genuinely believe their approach qualifies. But there's an enormous gap between:

"We use AI to analyze campaign data and provide recommendations."

and

"We use multi-LLM orchestration with human-in-the-loop validation to extract dimensional patterns from campaign performance while maintaining a privacy-first architecture."

The first statement could describe almost anything. The second describes something specific and sophisticated.

Your job is to distinguish between them. Because if you can't tell the difference, you'll end up paying for pattern recognition and receiving benchmark aggregation with better marketing.

Red Flags—What to Avoid

When evaluating campaign intelligence approaches, certain characteristics should immediately raise skepticism. They're not dealbreakers, necessarily—but they are signals that demand much deeper questioning.

Red Flag #1: Benchmarking Platforms That Claim Pattern Recognition

What it looks like: Industry benchmark providers adding "pattern recognition" or "AI-powered insights" to their existing aggregation platforms.

Why it matters: We spent the first three chapters of this book explaining why benchmark aggregation fails. Adding AI to the aggregation process doesn't fix the fundamental problem—you're still averaging incomparable campaigns and pretending the result is meaningful.

Real pattern recognition requires dimensional analysis across specific campaign characteristics. If the system is still organizing data primarily by industry vertical, company size, or channel type, it's not doing dimensional pattern matching—it's doing categorization with fancier labels.

The diagnostic question: "Can you show me campaigns that rhyme across different industries but similar dimensional profiles?" If they can't, they're not doing pattern recognition.

Red Flag #2: Off-the-Shelf AI Tools Claiming Sophisticated Analysis

What it looks like: Vendors offering access to ChatGPT, Claude, or other LLMs through simple interfaces, claiming this delivers pattern recognition capabilities.

Why it matters: Consumer-grade AI tools are remarkable. They're also available to everyone. If your "competitive advantage" is using Jasper or Copy.ai or any other off-the-shelf tool, you have no competitive advantage. Your competitors can access the exact same capabilities with a credit card.

Pattern recognition requires purpose-built infrastructure—embedding models for dimensional representation, similarity algorithms for pattern matching, orchestration frameworks for multi-LLM analysis, and validation processes for ensuring accuracy. None of this exists in off-the-shelf tools.

Off-the-shelf tools are valuable for many things. Pattern recognition at scale isn't one of them.

And there's an even more common deception: many vendors claim "proprietary AI" when what they actually mean is "we wrote some custom prompts." Prompt engineering is valuable—I use sophisticated prompting extensively. But templatized prompts for publicly available models aren't proprietary AI infrastructure. They're standardized usage of commodity tools. The distinction matters because one represents a defensible competitive advantage and the other is easily replicated by anyone with access to the same models.

The diagnostic question: "What proprietary infrastructure have you built specifically for dimensional pattern analysis?" If the answer is "We use [named AI tool]" or "we have proprietary prompts," that's a red flag.

Red Flag #3: Black Box Systems with Unexplainable Recommendations

What it looks like: "Our AI analyzed your campaign and recommends these changes"—without explaining the dimensional reasoning or showing comparable patterns.

Why it matters: Pattern recognition is valuable precisely because it's explainable. You should understand *why* a recommendation makes sense based on what happened in similar dimensional contexts. If the system can't articulate the pattern it identified, the dimensional similarities it detected, or the comparable outcomes that inform the recommendation, it's not doing pattern analysis—it's generating suggestions without clear reasoning.

This is particularly dangerous because it creates a false sense of sophistication. "The AI says do this" sounds authoritative. But if you can't evaluate whether the reasoning makes sense, you can't distinguish between genuine insight and statistical noise that happened to pass a significance threshold.

The diagnostic question: "Walk me through the dimensional reasoning behind that recommendation. What campaigns rhyme with mine, and why?" If they can't provide clear answers, that's a problem.

Red Flag #4: Single-LLM Approaches Without Orchestration

What it looks like: Vendors locked into one AI model—GPT-only, Claude-only, Gemini-only—without multi-model orchestration capabilities.

Why it matters: Different LLMs have different strengths, and those strengths shift as models evolve. We explored this in depth in Chapter 8—sophisticated pattern recognition leverages complementary capabilities across multiple models through orchestration, using the best tool for each specific analytical task.

Single-model approaches suggest either technical naivety (they don't understand the orchestration advantage) or infrastructure limitations (they can't build it). Either way, it means you're getting capabilities constrained by one model's limitations rather than benefits enhanced by multiple models' strengths.

The diagnostic question: "What models do you use, and how do you orchestrate them for different analytical tasks?" If the answer is one model for everything, that's limiting.

Red Flag #5: Technology-First Positioning

What it looks like: Marketing materials that lead with "our AI" or "our proprietary algorithm" rather than domain expertise and strategic judgment.

Why it matters: This reveals a fundamental misunderstanding of what makes pattern recognition valuable. The technology enables analysis at scale. But the intelligence comes from understanding which dimensions matter, what patterns are meaningful, how to validate AI outputs, and when strategic judgment should override algorithmic suggestions.

Vendors who lead with technology typically lack the domain expertise that makes technology useful. They've built sophisticated tools without understanding the sophisticated judgment required to use them well.

Remember Chapter 7's central argument: as AI capabilities advance and commoditize, experienced human judgment becomes the differentiator. If a vendor positions technology as their primary value, they're building on a depreciating asset rather than an appreciating one.

The diagnostic question: "Before we discuss your technology, tell me about your team's background in B2B marketing and campaign strategy." If they pivot back to technology without addressing expertise depth, that's revealing.

Red Flag #6: No Emphasis on Human-in-the-Loop Validation

What it looks like: Suggestions that AI handles pattern recognition automatically, with minimal human involvement required.

Why it matters: AI is remarkable at finding patterns. It's also remarkable at finding spurious correlations, overfitting to noise, and generating recommendations that are technically accurate but strategically nonsensical.

Human-in-the-loop processes aren't a limitation of current AI that will eventually be overcome. They're a fundamental requirement for ensuring that pattern recognition produces strategically sound recommendations rather than mathematically interesting but practically useless correlations.

Vendors who downplay HITL processes either don't understand this requirement or can't afford the experienced humans necessary to implement it well. Either way, you'll get unvalidated AI outputs that may or may not reflect genuine intelligence.

The diagnostic question: "What happens when your AI suggests a pattern that doesn't make strategic sense?" If they can't describe robust validation workflows, that's concerning.

Red Flag #7: Vague Privacy Architecture Explanations

What it looks like: General reassurances about data security without specific technical explanations of how competitive intelligence and client confidentiality are protected.

Why it matters: Chapter 9 established why privacy-first architecture isn't optional—it's fundamental to creating intelligence that doesn't compromise competitive advantage. If a vendor can't clearly explain their dimensional abstraction approach, what gets captured versus what gets protected, and how anonymization works at the technical level, they likely haven't built privacy into the architecture.

They're probably using access controls to protect data they've captured—which is policy-based privacy, not architecture-based privacy. And policy-based privacy fails when policies change, access controls are misconfigured, or human error happens.

The diagnostic question: "Explain your technical approach to privacy. How do you extract dimensional patterns without capturing campaign specifics?" If the answer is vague or focuses on access controls rather than architectural design, be skeptical.

These seven red flags don't automatically disqualify a vendor. But they should trigger much deeper questioning. And if you see multiple red flags clustering together, that's a strong signal that the approach is superficial rather than sophisticated.

Green Flags—What Signals Real Capability

So what should you look for? What signals indicate a vendor or approach that can actually deliver sophisticated pattern recognition?

The inverse of red flags, certainly. But there's more nuance here than simple opposites.

Green Flag #1: Multi-Dimensional Framework Articulation

What it looks like: Clear, detailed explanation of what dimensions they analyze and why those specific dimensions matter for campaign performance.

Why it matters: This reveals whether they've thought deeply about what actually drives campaign similarity. Anyone can say "we analyze multiple dimensions." Sophisticated practitioners can articulate exactly which dimensions, how they're extracted, why they matter, and how they interact.

In Chapter 6, we explored five core dimensions: audience behavior, competitive environment, message strategy, execution approach, and business goals. A sophisticated vendor should be able to discuss their dimensional framework with similar depth—and explain why they chose those specific dimensions rather than others.

This doesn't mean they need to use the exact same framework. But they should have a framework that's comparably thoughtful and specific.

What to listen for: Specificity. Reasoning. Evidence of iteration and refinement based on learning what actually predicts campaign similarity. If they can't articulate their dimensional framework clearly, they probably don't have one.

Green Flag #2: Human Expertise Prominence

What it looks like: Marketing materials and conversations that emphasize domain expertise and strategic judgment before discussing technology capabilities.

Why it matters: This signals a proper understanding of where value comes from. The technology enables pattern recognition at scale. But the expertise determines which patterns matter, how to validate AI outputs, when to question algorithmic suggestions, and how to translate patterns into strategic recommendations.

Look for evidence of deep B2B marketing experience—not just general marketing background, but specific expertise in complex buyer journeys, enterprise sales cycles, and multi-touch campaign orchestration. Look for multi-disciplinary backgrounds spanning creative strategy, quantitative analysis, and channel expertise.

Here's a useful heuristic for thinking about the expertise-technology relationship:

A vendor with 20+ years of domain expertise who added AI capabilities three years ago is dramatically more valuable than a vendor with three years of domain expertise who spent 20+ years building AI systems. The former built AI to enhance expertise. The latter built expertise to sell AI.

But the optimal scenario is something different entirely: deep domain expertise built into a company that's AI-native from inception. Not bolting AI onto legacy processes, but architecting everything around AI capabilities from the beginning while being guided by experienced strategic judgment. That combination—decades of refined expertise with no legacy constraints holding back AI integration—produces something neither traditional agencies nor AI-first startups can easily replicate.

This combination is genuinely rare. Most established agencies have deep expertise but legacy constraints. Most AI startups have technical sophistication but shallow marketing experience. The few organizations that combined decades of marketing expertise with AI-native architecture from inception—building around dimensional pattern recognition before it had a trendy name—occupy a unique position.

What to listen for: When they talk about their team, do they lead with domain expertise or technical capabilities? Both matter. But sequence reveals priority. And do they show evidence of building AI-native operations rather than retrofitting AI onto traditional processes?

Green Flag #3: Multi-LLM Orchestration Capabilities

What it looks like: Evidence of sophisticated AI architecture that leverages multiple models for different analytical tasks, with clear reasoning about when to use which model.

Why it matters: This demonstrates technical sophistication and pragmatic understanding of AI capabilities. As we discussed in Chapter 8, different models excel at different tasks, and those capabilities evolve as models improve.

Vendors with orchestration capabilities can match analytical tasks to model strengths—using appropriate models for dimensional reasoning, creative hypothesis generation, performance analysis, and ensemble approaches for validation. This produces better results than any single model.

It also signals that the vendor is building genuine infrastructure rather than just using consumer tools. Orchestration requires purpose-built frameworks, not just API access.

What to listen for: Specific examples of how they use different models for different purposes. Evidence of experimentation and learning about model strengths. Discussion of ensemble approaches for validation. Recognition that model capabilities evolve and orchestration strategies adapt accordingly.

Green Flag #4: Transparent Methodology

What it looks like: Willingness and ability to explain how pattern identification works, how dimensional similarity is calculated, how validation occurs, and how recommendations are generated.

Why it matters: Transparency builds trust, but it also reveals depth of thinking. Vendors who can clearly explain their methodology have thought carefully about every step. Vendors who deflect with "proprietary algorithm" claims often lack the sophistication they're implying.

This doesn't mean they need to reveal implementation details that compromise their competitive advantage. But they should be able to explain the conceptual approach in enough depth that you can evaluate whether it makes sense.

You're looking for the ability to walk you through the following: How do you extract dimensional characteristics from campaign descriptions? How do you calculate similarity between campaigns? How do you validate that the calculated similarity actually predicts comparable performance? How do you refine the dimensional framework based on new learning?

What to listen for: Clarity without mystification. Depth without deflection. Evidence of iterative refinement based on learning what works versus what seemed like it should work.

Green Flag #5: Privacy-First Architecture

What it looks like: Clear, specific technical explanation of how dimensional patterns are extracted without capturing campaign specifics, how anonymization works at the architectural level, and what protections ensure competitive intelligence stays protected.

Why it matters: Chapter 9 established why this matters, both ethically and practically. But it's also a signal of technical sophistication. Building privacy into architecture rather than protecting data through access controls requires a more sophisticated design. It's harder to build. Which means vendors who've actually done it have invested in getting this right.

Look for explanations that distinguish between what gets extracted (dimensional patterns and aggregate outcomes) versus what stays protected (campaign specifics, creative details, competitive intelligence). Look for discussion of architectural privacy versus policy privacy. Look for clarity about why dimensional abstraction produces better insights while simultaneously protecting competitive advantage.

What to listen for: Technical specificity. Distinction between extraction and protection. Understanding of why privacy constraints actually improve intelligence quality by forcing abstraction to transferable patterns.

Green Flag #6: HITL Process Emphasis

What it looks like: Detailed description of how experienced humans validate AI outputs, question recommendations that seem off, refine dimensional analysis, and exercise strategic judgment throughout the process.

Why it matters: This reveals an understanding that AI is a tool that amplifies human expertise rather than replaces it. Vendors who emphasize humans-in-the-loop have structured their operations around human judgment rather than around automation.

Look for specific workflows: Who reviews AI-extracted dimensional characteristics? How do they validate that the AI correctly identified relevant patterns? What happens when AI recommendations conflict with strategic judgment? How do insights get refined based on human expertise?

The sophistication of HITL processes often separates genuinely valuable pattern recognition from superficial analysis that happens to use AI.

What to listen for: Specific roles and responsibilities. Evidence of structured workflows. Examples of times human judgment overrode algorithmic suggestions. Discussion of how expertise improves AI outputs rather than just verifying them.

Green Flag #7: Domain Expertise Depth

What it looks like: Evidence of extensive B2B marketing experience—not just general background, but specific expertise in the contexts where pattern recognition adds most value.

Why it matters: Pattern recognition requires understanding what dimensions matter, what patterns are meaningful, what strategic contexts affect performance, and when conventional wisdom is wrong. This comes from years of experience seeing what actually works across diverse contexts.

Look for depth rather than breadth. Someone who's spent 20 years in B2B enterprise marketing, worked across industries, managed both successful and failed campaigns, and developed intuition about what drives performance is dramatically more valuable than someone with five years of marketing experience plus credentials in data science.

The dimensional frameworks that make pattern recognition valuable come from this deep expertise. The ability to validate AI outputs depends on this judgment. The strategic recommendations that turn patterns into action require this background.

What to listen for: Specific examples drawn from real experience. Evidence of learning from both successes and failures. Sophisticated understanding of contextual factors that influence campaign performance. Nuanced judgment about when patterns transfer versus when context matters more.

Green Flag #8: Strategic Judgment Emphasis

What it looks like: Discussion framed around "we help you decide what to test" rather than "our AI decides for you."

Why it matters: This reveals a proper understanding of how pattern intelligence gets used. The patterns inform strategic judgment—they don't replace it. You still need to decide whether a pattern that worked in similar contexts makes sense for your specific situation, whether to test the hypothesis the pattern suggests, and how to adapt the approach for your unique competitive position.

Vendors who position themselves as decision-makers—"do what our AI recommends"—are either naive about strategic complexity or deliberately positioning AI as having more authority than it deserves. Either way, that's problematic.

Sophisticated vendors position themselves as strategic advisors who bring pattern intelligence to inform your decisions—not as oracles whose recommendations you should follow automatically.

What to listen for: Language of collaboration and exploration rather than prescription and certainty. Emphasis on testing hypotheses rather than implementing proven solutions. Recognition that you understand your specific context better than any AI possibly could.

Green Flag #9: Performance + Pattern Learning Framework

What it looks like: Discussion of success that includes both quantitative outcomes *and* qualitative learning that compounds across campaigns.

Why it matters: Here's a reality we can't ignore: every marketer needs to see performance metrics. You're evaluated on open rates, conversion rates, pipeline contribution, and ROI. Pattern recognition doesn't eliminate that need—nor should it.

But here's what pattern recognition changes: it provides the context that makes those metrics meaningful.

Traditional approach: "Our email campaign achieved a 22% open rate." In comparison to what? Last month's campaign? Industry benchmark? Is that good or bad? Should you do more of this or less?

Pattern recognition approach: "Our email campaign achieved a 22% open rate, which is 28% above what similar campaigns targeting risk-averse executives in competitive markets typically achieve with feature-focused messaging. This validates that trust-building message strategy outperforms in this dimensional context—a pattern we can now apply across six other active campaigns with similar audience and market characteristics."

The quantitative result is the same. But the qualitative learning compounds.

You still measure performance. But you also extract transferable intelligence that makes future campaigns smarter. You're not just optimizing individual campaigns—you're building institutional knowledge that appreciates over time.

Sophisticated vendors understand this dual value. They'll discuss campaign performance *and* pattern learning. They'll show you both "what happened" and "what we learned that applies elsewhere."

Vendors who only talk about performance metrics are stuck in the benchmark mindset—evaluating campaigns in isolation. Vendors who only talk about learning without tying it to performance are disconnected from business reality.

Look for vendors who integrate both naturally—because that's how pattern recognition actually creates value.

What to listen for: Discussion of how insights transfer across campaigns. Examples of patterns learned from one campaign that improved performance in others. Recognition that quantitative performance is essential *and* that qualitative learning compounds value over time.

Green Flag #10: Collaborative Partnership Model

What it looks like: Vendors who actively solicit your challenging questions, welcome strategic input, and position themselves as thinking partners rather than service providers.

Why it matters: The best pattern recognition relationships don't feel like you're being presented *to*—they feel like you're thinking *together*.

Throughout my career, I've found that my most successful client relationships are the ones that operate like true partnerships. Clients who seek understanding, challenge perspectives, and ask hard questions push me to deliver better work. When clients are engaged and participating actively in the strategic process, we arrive at better solutions than either party would have reached independently.

Yet many agencies try to avoid this dynamic. They minimize client involvement, try to hide problems, sweep challenges under the rug, and position themselves as the experts who just need you to approve their recommendations. This creates a transactional relationship that limits what's possible.

Look for vendors who explicitly welcome your involvement. Who want to understand your strategic context deeply. Who see your questions as opportunities to refine thinking rather than challenges to defend against. Who bring you into the pattern analysis process rather than just presenting conclusions.

This collaborative approach connects directly to the human-in-the-loop principle—but extends it to include *your* expertise and judgment in the loop. Pattern recognition works best when it combines the vendor's pattern intelligence with your deep knowledge of your specific market position, competitive dynamics, and strategic priorities.

What to listen for: An invitation to engage deeply rather than just approve recommendations. Questions about your strategic context that go beyond surface details. A willingness to explain their reasoning and adapt based on your input. Evidence of partnerships where client involvement made the work stronger.

These green flags cluster in meaningful ways. Vendors who lead with human expertise typically also emphasize HITL validation and strategic judgment. Vendors with transparent methodology usually also have a clear privacy architecture. Vendors with multi-LLM orchestration often have sophisticated dimensional frameworks. Vendors who emphasize performance and learning tend to welcome collaborative partnerships.

Look for patterns in the green flags themselves.

Questions That Separate Signal from Noise

When evaluating potential partners or approaches, certain questions cut through marketing language to reveal actual capability. These aren't designed as gotcha questions. They're diagnostic questions that expose depth—or lack thereof.

On Dimensional Analysis

"Walk me through how you identify campaigns that rhyme. What dimensions do you analyze, and why those specific dimensions?"

Sophisticated answers will be specific and detailed. They'll explain not just what dimensions they use, but why those dimensions matter, how they're weighted for different analytical purposes, and how the dimensional framework has evolved based on learning.

Superficial answers will be vague—"we analyze multiple dimensions like industry, size, and channel"—or will deflect to technology—"our AI identifies the relevant patterns automatically."

"Can you show me examples of campaigns that rhyme across different industries but similar dimensional profiles?"

This tests whether they actually do dimensional pattern matching or just categorical grouping. True pattern recognition should find similarities between, say, a healthcare SaaS campaign and a financial services campaign that share audience risk aversion, competitive intensity, and message strategy characteristics—even though their industries are completely different.

If they can't provide cross-industry examples, they're probably doing benchmark aggregation with better labels.

On Human Expertise

"Tell me about your team's background in B2B marketing before we discuss your technology."

This tests whether expertise or technology is primary. Sophisticated vendors will happily discuss team backgrounds, experience depth, and how that expertise shapes their approach. They'll talk about learning from campaigns, developing judgment over years, and using AI to scale that expertise.

Vendors who immediately pivot back to technology—"well, our AI is trained on thousands of campaigns"—are revealing that technology is their primary asset. Which means their primary asset is depreciating as AI capabilities commoditize.

"What happens when your AI suggests a pattern that doesn't make strategic sense?"

This reveals HITL sophistication. Good answers describe specific validation workflows, examples of times human judgment overrode AI recommendations, and how expertise refines algorithmic outputs.

Bad answers suggest AI is always right, human input is minimal, or validation is mostly automated.

On Privacy and Protection

"How do you protect my competitive advantage when learning from my campaign patterns?"

This tests whether they understand the dual challenge: creating collective intelligence while protecting individual advantage. Sophisticated answers distinguish between dimensional patterns (which transfer) and campaign specifics (which stay protected). They explain architectural privacy rather than just access controls.

Vague answers about "data security" or "confidentiality agreements" suggest they haven't thought deeply about this challenge.

"Can you explain your technical approach to anonymization? What gets extracted versus what stays protected?"

This tests architectural sophistication. Good answers are specific about dimensional abstraction, explain why certain details never get captured, and articulate how privacy constraints actually improve intelligence quality.

Deflection or vagueness here is a significant red flag.

On Methodology

"How do you validate that dimensional similarity actually predicts comparable performance patterns?"

This tests whether they're doing rigorous pattern recognition or just mathematical correlation. Sophisticated answers discuss validation frameworks, evidence that similar dimensional profiles produce similar outcome patterns, and iterative refinement based on learning.

Superficial answers appeal to authority—"our AI is very advanced"—without explaining actual validation.

"Walk me through how you refine your dimensional framework based on new learning."

This tests whether the approach is static or evolving. Pattern recognition should get better as it analyzes more campaigns. The dimensional framework should be refined as you learn which characteristics actually predict similarity versus which seemed like they should matter but don't.

Good answers provide specific examples of how the framework has evolved. Bad answers suggest the framework is fixed or that refinement happens automatically without human judgment.

On Strategic Application

"How do you help me decide whether a pattern that worked in similar contexts makes sense for my specific situation?"

This tests whether they position pattern intelligence as decision support or decision replacement. Sophisticated vendors will discuss collaboration, contextual factors, testing hypotheses, and adapting general patterns to specific situations.

Vendors who suggest their recommendations should be followed automatically don't understand the strategic complexity you're navigating.

"Can you walk me through examples of when human judgment should override pattern-based recommendations?"

This is a brilliant diagnostic because it tests humility and strategic sophistication simultaneously. Great vendors will provide thoughtful examples of contexts where patterns don't transfer, where strategic judgment matters more than historical similarity, or where unique competitive positions require departing from what worked elsewhere.

Vendors who can't provide good answers either lack the experience to recognize these situations or lack the humility to acknowledge AI limitations.

On Performance and Learning

"How do you measure success? What do I get beyond campaign performance metrics?"

This tests whether they understand the dual value of pattern recognition. Good answers discuss both quantitative outcomes (the campaign results) and qualitative learning (the transferable insights that improve future campaigns).

Superficial answers focus only on metrics—"we improve your open rates by X%"—without discussing how the learning compounds across campaigns.

"Can you show me examples of insights learned from one campaign that improved performance in another?"

This tests whether they actually extract and apply patterns across campaigns. Sophisticated vendors will have specific examples of how dimensional learning transferred from one context to another, creating compounding value.

If they can't provide these examples, they're doing campaign optimization, not pattern recognition.

These questions aren't designed to be asked robotically down a checklist. They're conversation starters that reveal depth through how vendors respond, what they emphasize, and where they struggle to provide specific answers.

Pay attention to the following:

- Specificity versus vagueness: Sophisticated vendors are specific. Superficial vendors are vague.

- Expertise versus technology emphasis: Where do they lead? What do they emphasize?

- Clarity versus mystification: Do they explain clearly or hide behind complexity?

- Examples versus claims: Do they illustrate with specific examples or just make general claims?

- Humility versus certainty: Do they acknowledge limitations or claim their approach always works?

- Partnership versus vendor framing: Do they position themselves as collaborators or service providers?

The pattern in the answers matters as much as any individual answer.

Why Most Vendors Will Fail These Tests

Most vendors claiming pattern recognition capabilities won't pass these evaluations.

Not because they're dishonest. Most genuinely believe they're doing sophisticated analysis. But there's a massive gap between:

- Using AI tools to analyze campaign data (which is relatively easy)

- Building purpose-built infrastructure for dimensional pattern recognition (which is hard)

And between:

- Having some marketing experience (which is common)
- Having deep multi-disciplinary B2B expertise refined over decades (which is rare)

Between:

- Implementing some human review of AI outputs (basic)
- Building sophisticated HITL workflows where expertise shapes every stage (advanced)

Between:

- Securing data through access controls (standard practice)
- Designing privacy into architecture through dimensional abstraction (sophisticated design)

Between:

- Optimizing individual campaigns (tactical)
- Extracting transferable patterns that compound across campaigns (strategic)

The vendors who can actually deliver sophisticated pattern recognition have invested years building expertise, developed purpose-built infrastructure, designed privacy into architecture, structured operations around human judgment amplified by AI, and built collaborative partnership models that integrate client expertise into the analysis process.

That's not something you build casually. It's not something you achieve by adding AI features to existing platforms. It's not something you replicate by accessing consumer AI tools through better interfaces. It's not something you bolt onto traditional agency operations without fundamental architectural changes.

Which means genuine pattern recognition capability is rare—precisely because it's hard to build.

Your evaluation framework helps you identify the rare vendors who've actually done the work versus the many vendors who've adopted the language without building the capability.

The Bottom Line

Pattern recognition done well transforms campaign intelligence from generic benchmarks to genuinely useful insights. But "done well" is the critical qualifier.

The sophistication gap between superficial and genuine pattern recognition is enormous. The market is about to fill with vendors claiming capabilities they haven't built. And the marketing language will make them all sound plausible.

Your ability to evaluate approaches determines whether you get transformative intelligence or expensive benchmark aggregation with AI labels.

Use these red flags, green flags, and diagnostic questions not as rigid checklists but as frameworks for developing judgment. Pay attention to patterns in how vendors respond—where they're specific versus vague, where they emphasize expertise versus technology, where they demonstrate depth versus deflect with marketing language, and where they invite collaboration versus maintaining vendor distance.

The vendors who can actually deliver sophisticated pattern recognition will welcome these questions. They'll appreciate that you understand what makes the approach valuable. They'll engage substantively because they've built something genuine. They'll see your questions as the beginning of a collaborative partnership rather than as obstacles to closing a sale.

The vendors who struggle with these questions are revealing that their capability doesn't match their claims—which is valuable information before you commit resources.

Pattern recognition promises to transform how you optimize campaigns. But only if you can recognize which vendors deliver genuine patterns versus which vendors deliver noisy patterns.

Which, appropriately enough, requires developing pattern recognition about pattern recognition itself.

We've now established what pattern recognition is, why it requires human expertise, how AI enables it, why privacy-first architecture matters, and how to evaluate whether approaches are sophisticated or superficial.

You have the intellectual framework. You understand why benchmarks fail and what actually works. You can distinguish genuine pattern recognition from benchmark aggregation with AI labels.

But understanding and evaluation are only valuable if they lead to action.

The next question isn't whether pattern recognition will replace benchmarks—that transition is already beginning among sophisticated marketers who've grown tired of comparing campaigns to statistical fiction. The question is how you'll participate in that transition.

Because there's no universal path forward. What makes sense for a global agency running hundreds of campaigns annually is different from what makes sense for a regional consultancy running dozens, which is different from what makes sense for an in-house team running ten.

Some organizations should build internal capability. Others should partner with specialists who've already built it. Still others should start with dimensional thinking while determining their longer-term approach.

Chapter 11 provides the decision framework—helping you assess which path aligns with your situation, resources, and strategic priorities. Not to prescribe a single answer, but to help you make an informed choice about how to stop playing benchmark theater and start building genuine competitive advantages.

The evaluation capability you now possess is only powerful if it leads to a decision. Let's explore what your options actually are.

PART 3
THE FUTURE

Chapter Eleven

The Choice Ahead

EVERY SOPHISTICATED MARKETER HAS had that moment—staring at a benchmark report, knowing it's bullshit, but playing along because what else could you do? That silence is finally breaking.

Not because marketers suddenly discovered benchmarks are flawed. We've always known. We've just never had the vocabulary—or more importantly, the alternative—to articulate why averaging incomparable campaigns into industry benchmarks was theater rather than intelligence.

Now we do. Dimensional pattern recognition isn't just a better way to measure campaigns. It's what sophisticated marketers have been trying to approximate intuitively for years—finding campaigns that actually rhyme across dimensions that matter, not campaigns that happen to share surface categories.

The question isn't whether you've recognized the problem. If you've read this far, you've likely been frustrated with benchmarks for years. The question is what you do about it now that a genuine alternative exists.

There are three paths forward. Each has different requirements, economics, and strategic implications. Your situation determines which makes sense.

These paths are relevant whether you are an agency or a brand.

Path 1: Building Internal Capability

This means developing dimensional pattern recognition as an organizational competency—building the infrastructure, developing the expertise, and accumulating pattern intelligence over time.

Let's be explicit about what "building" means, because it's more complex than "hire some AI people and figure it out." You'll need to develop technical infrastructure, human expertise, and an operational commitment, and make determinations about a timeline and economic considerations:

Technical infrastructure

You need systems for embedding generation (converting campaign descriptions into mathematical representations), vector storage and retrieval (enabling similarity search across high-dimensional spaces), pattern extraction workflows, multi-LLM orchestration, and privacy-preserving architecture. This isn't a one-time build—it requires continuous maintenance as technology evolves and new capabilities emerge.

Human expertise

You need practitioners with both sophisticated AI orchestration capabilities *and* deep multi-disciplinary marketing experience, and they can be hard to find. Not specialists who've only worked in SaaS, or only in healthcare, or only in enterprise. The pattern recognition insight that "this cybersecurity campaign rhymes with that healthcare compliance campaign despite having no industry overlap" comes from humans who've actually run campaigns in both contexts and recognize structural similarities.

This expertise is genuinely rare. Most marketing careers naturally specialize. Finding practitioners with breadth across enterprise software, financial services, healthcare, manufacturing, and professional services—and the judgment to recognize which patterns transfer—commands premium compensation.

THE CHOICE AHEAD

Operational commitment

Every campaign needs to be treated as an intelligence-gathering opportunity. Dimensional analysis before launch. Pattern-informed hypothesis generation. Performance tracking against predictions. Retrospective analysis extracting learnings. Pattern library updates incorporating new insights. This requires operational discipline and strategic commitment—not just technology deployment.

Time to proficiency

You'll need to invest 18-24 months minimum to approach sophisticated capability. Why? Because pattern recognition expertise comes from analyzing hundreds of campaigns across diverse contexts, encountering edge cases, refining judgment about what transfers, and accumulating institutional memory. You can build technical infrastructure in 3-6 months. But developing sophisticated pattern recognition fluency requires extensive campaign exposure over time.

The economics

- Personnel: $400K-800K annually for minimum viable team

- Infrastructure: $3K-14K monthly for cloud, AI APIs, tools

- Opportunity cost: 18-24 months of campaigns run without pattern intelligence

- Break-even: Typically requires 30+ campaigns annually to justify

When building makes sense

- High campaign volume (30+ annually) provides sufficient learning opportunities

- Strategic imperative for proprietary capability

- Long-term commitment to dimensional intelligence as core competency

- Budget for 18-24 month development before proficiency

- Ability to hire/develop rare talent combining AI expertise with multi-disciplinary marketing depth

If you have all of these conditions, building can create defensible competitive advantage. If you're missing several, partnership likely provides better outcomes faster.

Path 2: Partnering with Specialists

This means accessing established pattern recognition capability through collaboration with firms that have already built sophisticated infrastructure and accumulated dimensional intelligence.

Partnership with specialists provides immediate access to their pattern libraries and other resources, and the benefits of their proven expertise:

Immediate capability access

Partners who've invested years accumulating dimensional intelligence bring pattern libraries you can't replicate quickly; cross-industry insights that inform your vertical; dimensionally similar campaigns that reveal non-obvious strategic opportunities; validated patterns showing what actually transfers versus what sounds good but doesn't work; and edge case awareness from encountering failures across diverse contexts.

Proven expertise

Partners with multi-disciplinary teams bring strategic depth that's rare to assemble internally. Practitioners who've run campaigns across multiple industries have recognized dimensional patterns before pattern recognition had a name. They have calibrated judgment about when patterns transfer and when context overrides and quality control capabilities for catching subtle strategic errors in sophisticated AI outputs.

THE CHOICE AHEAD

There are two main options when it comes to how to structure your partnership:

White-label partnership

The specialist's capabilities are presented as your own. Your clients see sophisticated AI-powered pattern recognition without knowing about the partnership. This works when you want to strengthen competitive positioning without introducing additional complexity into client relationships.

Transparent partnership

The collaboration is explicit. You position partnership as strategic resource allocation—focusing on core competencies while accessing specialized capability. This works when clients value honesty about how you deliver sophistication.

There are also several economic structures to consider:

- Campaign-based: $5K–25K per campaign, depending on complexity
- Retainer model: $10K–40K monthly for ongoing access
- No permanent overhead during low-activity periods
- Immediate access without 18–24-month development
- No talent recruitment and retention risk

Partnering makes sense when the following conditions apply:

- Campaign volume of 10–30 annually (substantial but not enough for build economics)
- Need immediate capability without a development timeline
- Prefer to focus resources on core competencies
- Value cross-client pattern intelligence
- Recognize talent acquisition challenges for AI + marketing expertise combination

Partnership isn't admitting you can't build it. It's strategic resource allocation—accessing specialized capability that takes years to develop while focusing internal investment where you have a genuine competitive advantage.

Path 3: Starting with Dimensional Thinking

This means adopting the conceptual framework to improve strategic decision-making without infrastructure investment.

You map campaigns manually across the five dimensions. You identify dimensionally similar campaigns through deliberate analysis rather than computational scale. You challenge benchmark-dependent thinking in strategic conversations. You build fluency with dimensional concepts that inform better decisions even without systematic pattern recognition.

This approach doesn't scale. But it builds strategic thinking capability immediately. You can't analyze hundreds of campaigns computationally, but you can recognize when your cybersecurity campaign rhymes dimensionally with a healthcare compliance campaign you ran last year. That recognition—even without sophisticated infrastructure—improves strategic decisions.

There are several circumstances where this approach makes the most sense:

- You're an individual contributor without organizational authority.

- You're testing concepts before you're ready to commit to a partnership or infrastructure investment.

- You're building internal advocacy for future investment.

- Your company's campaign volume is too low (<10 annually) for other approaches.

This isn't a permanent solution. But it's a legitimate starting point that builds fluency with concepts that matter regardless of future build-or-partner decisions.

The Decision Framework

Before deciding which of the three paths to take, you'll need to perform an honest assessment of several factors:

What's your strategic priority?

If dimensional pattern intelligence is central to your competitive differentiation strategy, and you have resources to build properly, build. If it strengthens capabilities but isn't a core competency, partner. If you're exploring whether this creates meaningful value, start with dimensional thinking.

What's your realistic timeline?

If you need sophisticated capability this quarter, partner. If you can invest 18–24 months developing capability, consider building. If you're in exploration mode without immediate pressure, start with dimensional thinking.

What's your honest capability assessment?

Can you attract rare AI + multi-disciplinary marketing expertise? Have you successfully built technical capabilities before? Do you have organizational commitment to an 18–24-month development period? If yes to all, build. If no to several, partner.

What's your resource allocation philosophy?

Does building proprietary capability in every strategic domain align with your philosophy and capital availability? Or does partnering for capabilities outside core competency make more strategic sense?

There's no universal right answer. The right path depends on your specific context, resources, constraints, and strategic priorities.

The First-Mover Opportunity

Here's what makes timing critical regardless of which path you choose:

Right now, we're in a unique moment. The technology to enable dimensional pattern recognition at scale has emerged in the last 2–3 years. The frameworks to make it strategically useful exist. But the number of organizations that have built genuine capability—not just AI-powered benchmark aggregation—can be counted on one hand.

This won't last. By 2028, every agency and consultancy will claim pattern recognition capabilities. Most will be repackaging the same benchmark aggregation with better marketing. But a few will have built genuine dimensional intelligence over those three years.

The competitive advantage doesn't come from being faster than everyone else at adopting commodity capability. It comes from recognizing genuine capability before the market gets flooded with imitators.

First movers have a lot to gain:

Pattern intelligence accumulation

Every campaign you analyze dimensionally adds to pattern intelligence that improves future campaigns. Start now, and by 2028, you'll have two + years of accumulated intelligence. Start in 2028, and you'll be building from scratch while others apply sophisticated patterns.

Market definition authority

The firms that establish pattern recognition capability first will define how the market understands it. They'll establish the vocabulary, set the standards, and shape expectations. They'll be the ones others are compared against.

Talent attraction advantage

Sophisticated practitioners want to work where pattern intelligence is most advanced. Early capability development creates a talent magnet that strengthens over time.

Client relationship depth

Clients who experience genuine dimensional insights—discovering their campaigns rhyme with patterns from completely different industries—develop different relationships with partners who provide that intelligence. It's not vendor management. It's strategic collaboration.

The first-mover advantage isn't about being first to use new technology. It's about being among the few who built genuine capability before everyone started claiming they had it.

The Scarcity Reality

Let's be direct about current market reality: almost no one is actually doing sophisticated dimensional pattern recognition today.

Yes, many vendors claim "AI-powered pattern recognition." They're mostly doing benchmark aggregation with better marketing. They use ChatGPT to generate insights. They average campaigns within slightly narrower categories. They add AI labels to traditional analysis.

But genuine dimensional pattern recognition—multi-LLM orchestration extracting patterns across dozens of dimensions, with human expertise validating strategic relevance, building pattern libraries that compound over time? That capability is genuinely rare.

This scarcity is your opportunity. You're not catching up to a market that's already transformed. You're potentially among the first to recognize that transformation is becoming possible—and to act on that recognition before it becomes obvious.

By moving now, you're not following a trend. You're anticipating an evolution that sophisticated marketers have wanted for years but that only recently became technically feasible.

The Platform AI Trap

There's another dimension to consider as you evaluate your path forward—the one that Mike O'Toole crystallized perfectly at that industry dinner I mentioned in Chapter 7, about how platform AI's optimization recommendations are aligned with their shareholder value, not your campaign outcomes.

Every major platform now offers AI-powered optimization. LinkedIn's Campaign Manager has AI recommendations. Meta has Advantage+ campaigns. Google has Performance Max. They promise to handle the complexity for you—just feed them a budget and let their AI optimize.

But sophisticated marketers understand something important: platform AI optimizes for their platform revenue, not your business outcomes.

When Meta's AI suggests expanding your audience or increasing your budget by 40%, that recommendation comes from algorithms trained to maximize Meta's revenue. When Google's Performance Max shifts budget from search to YouTube, that's Google's AI pursuing Google's interests. These platforms will never tell you to spend less. They'll never suggest you'd get better results elsewhere.

You're essentially letting your media vendors determine your strategy.

This creates a compound trap when combined with benchmark dependence:

- Benchmarks tell you what "average" looks like (meaningless)

- Platform AI optimizes toward platform goals (misaligned)

- You're left with neither genuine intelligence nor objective optimization

The firms that break free from *both* traps—benchmark theater *and* platform AI dependency—are the ones positioned to build genuine competitive advantages.

Dimensional pattern recognition is fundamentally different. It's objective intelligence with no stake in where you allocate budget. It might tell you, "Campaigns with your dimensional profile typically see 3x better performance moving budget from Meta to LinkedIn" or "Your audience psychology suggests email will outperform display regardless of platform sophistication."

That's intelligence you'll never get from platform AI, because platform AI is structurally incapable of recommending against platform interests.

A Note on Evaluation

If you're considering partnering rather than building, Chapter 10's evaluation framework becomes critical. Most vendors claiming pattern recognition capability don't actually have it. They've recognized the market need and adjusted their marketing language, but they haven't built the infrastructure, accumulated the pattern intelligence, or developed the expertise.

Look for evidence of dimensional thinking that predates the current AI hype. Look for practitioners with multi-disciplinary experience across diverse industries. Look for specific explanations of how dimensional patterns are extracted and validated. Look for privacy-first architecture that protects competitive advantage while enabling pattern intelligence.

The vendors who can actually deliver sophisticated pattern recognition won't be defensive about these questions. They'll welcome them—because they've built something genuine and they recognize you understand what makes it valuable.

The Choice Is Yours

You've spent ten chapters understanding why benchmarks fail mathematically and psychologically, how dimensional pattern recognition works conceptually, why human expertise matters as AI advances, what technical architecture enables pattern intelligence, and how privacy protection ensures competitive advantage isn't compromised.

The intellectual argument is complete. The strategic framework is clear. The evaluation criteria are established.

Now you face one decision: act on what you've learned, or continue participating in benchmark theater while knowing it's meaningless.

This isn't about fear of being left behind. It's about the recognition of opportunity. The capability to do sophisticated dimensional pattern recognition has only become technically feasible in the last 2–3 years. The number of organizations that have built genuine capability remains tiny. The window to establish first-mover advantage is open, but it won't remain so.

Every campaign you run, starting tomorrow, either contributes to pattern intelligence or it doesn't. If you start building dimensional intelligence now—whether internally or through partnership—every campaign becomes smarter than the last. If you wait, those campaigns and their patterns are lost permanently.

Pattern intelligence compounds from the day you start. There's no way to accelerate that accumulation later. Time is the irreplaceable input.

The benchmark lie has been exposed. The alternative has been explained. The path forward is clear.

What happens next is up to you.

A Final Word

IF YOU'VE MADE IT this far, you've just invested several hours reading about why the marketing measurement tools you've relied on for years are fundamentally broken. That's not a comfortable realization.

I know because I spent 25+ years believing the same lie.

The lie isn't that benchmarks are real. *The lie is that they mean something.*

The lie is that averages provide intelligence. That comparing your campaign to thousands of incomparable campaigns tells you something useful about what to do next. That beating an industry benchmark means you succeeded, or missing it means you failed.

The lie is mathematical. Industry benchmarks aggregate campaigns with completely different audience behaviors, competitive pressures, decision complexities, organizational dynamics, and success definitions—then pretend the resulting average provides strategic guidance. It's statistical fiction dressed up as marketing intelligence.

The lie is psychological. We cling to benchmarks not because they work, but because they're comfortable. They let us celebrate "wins" that aren't really wins, explain away "losses" that mask opportunities, and avoid the harder work of understanding what actually drives performance in our specific context.

The lie is pervasive. Every marketing platform publishes benchmark reports. Every agency references industry standards. Every CMO presentation includes competitive comparisons to category averages. We've built an entire infrastructure around meaningless numbers, and everyone pretends it's sophisticated because admitting otherwise is too uncomfortable.

But here's the thing about lies: once you see them clearly, you can't unsee them.

You can probably tell that replicating the thinking and methodology discussed in this book takes tremendous effort. In fact, I've been working on it for years now—I even trademarked a name for it. I'm not going to hand you DIY worksheets or reveal the technical implementation details. That's not how defensible competitive advantages work.

What I have done is expose the lie completely, and provided you with a viable alternative. You now understand why benchmarks fail mathematically. You also know how dimensional pattern recognition works conceptually, why it requires experienced humans orchestrating AI rather than AI running on autopilot, and why this creates advantages that appreciate over time instead of commoditizing.

You have a choice ahead of you. You can keep believing the lie—comparing your campaigns to meaningless averages, wondering why that competitor who "doesn't know what they're doing" keeps beating you, trying to explain to your CEO why you hit the industry benchmark but still didn't hit your revenue target.

Or you can stop measuring success against statistical fiction and start building real intelligence based on genuine patterns.

Here's what's inevitable: the benchmark lie is collapsing. Not because of this book, but because AI makes sophisticated pattern recognition practically possible at scale—and once that capability exists, benchmark-dependent approaches become obviously obsolete.

Navigation evolved from paper maps to GPS to real-time traffic optimization, whether individual drivers adopted it or not. The people who led each transition got somewhere faster. The people who followed got somewhere eventually. The people who insisted paper maps were fine got really good at pulling over to figure out where they went wrong.

You can't stop the shift by waiting to see if it's real. You can only choose whether you lead it or explain later why you waited.

If you want to talk about what dimensional pattern recognition looks like for your specific situation—whether you're an agency trying to differentiate in an AI-saturated market, a brand trying to figure out which AI claims are real versus marketing theater, or just someone who read this and thought "finally, someone said what I've been thinking"—reach out.

A FINAL WORD

I'm not interested in convincing anyone who isn't already convinced by the argument. This book did that work. But if you've seen through the lie and want to build advantages based on genuine intelligence instead of comfortable fiction, let's have that conversation.

The benchmark lie has been exposed. What you do with that knowledge is up to you.

Jeffrey Porzio
Founder & Principal, Cadence B2B
jporzio@cadenceb2b.com
cadenceb2b.com

www.ingramcontent.com/pod-product-compliance
Lightning Source LLC
LaVergne TN
LVHW010317070526
838199LV00065B/5594